2D Materials for Biomedical Applications

2D Materials for Biomedical Applications

Editors

Minas M. Stylianakis
Athanasios Skouras

Basel • Beijing • Wuhan • Barcelona • Belgrade • Novi Sad • Cluj • Manchester

Editors
Minas M. Stylianakis
Department of Nursing
Hellenic Mediterranean
University
Heraklion
Greece

Athanasios Skouras
Department of Nursing
Hellenic Mediterranean
University
Heraklion
Greece

Editorial Office
MDPI
St. Alban-Anlage 66
4052 Basel, Switzerland

This is a reprint of articles from the Special Issue published online in the open access journal *Molecules* (ISSN 1420-3049) (available at: www.mdpi.com/journal/molecules/special_issues/2D_materials_biomedical).

For citation purposes, cite each article independently as indicated on the article page online and as indicated below:

Lastname, A.A.; Lastname, B.B. Article Title. *Journal Name* **Year**, *Volume Number*, Page Range.

ISBN 978-3-0365-8739-4 (Hbk)
ISBN 978-3-0365-8738-7 (PDF)
doi.org/10.3390/books978-3-0365-8738-7

© 2023 by the authors. Articles in this book are Open Access and distributed under the Creative Commons Attribution (CC BY) license. The book as a whole is distributed by MDPI under the terms and conditions of the Creative Commons Attribution-NonCommercial-NoDerivs (CC BY-NC-ND) license.

Contents

About the Editors vii

Preface ix

Vladislav V. Shunaev and Olga E. Glukhova
Nanoindentation of Graphene/Phospholipid Nanocomposite: A Molecular Dynamics Study
Reprinted from: *Molecules* 2021, 26, 346, doi:10.3390/molecules26020346 1

Pengyu Gong, Yi Zhou, Hui Li, Jie Zhang, Yuying Wu and Peiru Zheng et al.
Theoretical Study on the Aggregation and Adsorption Behaviors of Anticancer Drug Molecules on Graphene/Graphene Oxide Surface
Reprinted from: *Molecules* 2022, 27, 6742, doi:10.3390/molecules27196742 10

Mohammad Oves, Mohammad Omaish Ansari, Mohammad Shahnawaze Ansari and Adnan Memić
Graphene@Curcumin-Copper Paintable Coatings for the Prevention of Nosocomial Microbial Infection
Reprinted from: *Molecules* 2023, 28, 2814, doi:10.3390/molecules28062814 22

Nayara Balaba, Silvia Jaerger, Dienifer F. L. Horsth, Julia de O. Primo, Jamille de S. Correa and Carla Bittencourt et al.
Polysaccharides as Green Fuels for the Synthesis of MgO: Characterization and Evaluation of Antimicrobial Activities
Reprinted from: *Molecules* 2022, 28, 142, doi:10.3390/molecules28010142 36

Kanae Suzuki, Misato Iwatsu, Takayuki Mokudai, Maiko Furuya, Kotone Yokota and Hiroyasu Kanetaka et al.
Visible-Light-Enhanced Antibacterial Activity of Silver and Copper Co-Doped Titania Formed on Titanium via Chemical and Thermal Treatments
Reprinted from: *Molecules* 2023, 28, 650, doi:10.3390/molecules28020650 50

Muhammad Dilshad, Afzal Shah and Shamsa Munir
Electroanalysis of Ibuprofen and Its Interaction with Bovine Serum Albumin
Reprinted from: *Molecules* 2022, 28, 49, doi:10.3390/molecules28010049 60

Michele Massa, Mirko Rivara, Thelma A. Pertinhez, Carlotta Compari, Gaetano Donofrio and Luigi Cristofolini et al.
Chemico-Physical Properties of Some 1,1′-Bis-alkyl-2,2′-hexane-1,6-diyl-bispyridinium Chlorides Hydrogenated and Partially Fluorinated for Gene Delivery
Reprinted from: *Molecules* 2023, 28, 3585, doi:10.3390/molecules28083585 78

Sergey Lazarev, Sofya Uzhviyuk, Mikhail Rayev, Valeria Timganova, Maria Bochkova and Olga Khaziakhmatova et al.
Interaction of Graphene Oxide Nanoparticles with Human Mesenchymal Stem Cells Visualized in the Cell-IQ System
Reprinted from: *Molecules* 2023, 28, 4148, doi:10.3390/molecules28104148 95

Qi Zhang, Fengjiao Xu, Pei Lu, Di Zhu, Lihui Yuwen and Lianhui Wang
Efficient Preparation of Small-Sized Transition Metal Dichalcogenide Nanosheets by Polymer-Assisted Ball Milling
Reprinted from: *Molecules* 2022, 27, 7810, doi:10.3390/molecules27227810 110

Faten Bashar Kamal Eddin and Yap Wing Fen
The Principle of Nanomaterials Based Surface Plasmon Resonance Biosensors and Its Potential for Dopamine Detection
Reprinted from: *Molecules* **2020**, *25*, 2769, doi:10.3390/molecules25122769 . . . . **122**

About the Editors

Minas M. Stylianakis

Minas M. Stylianakis is an assistant professor at the Department of Nursing of the Hellenic Mediterranean University and a member of the Hybrid Nanostructures group of the Institute of Electronic Structure and Laser (IESL) of FORTH. He received his Ph.D. degree in Chemistry from the University of Crete in 2015. His expertise lies in the synthesis, solution processing, and characterization of universal carbon- and graphene-based materials, 2D materials, small molecules, and polymers. His research interests include the development of biomedical and environmental applications, focusing on drug delivery systems and implants design/development, self-healing, and antimicrobial coatings, water treatment technologies, and energy production.

Athanasios Skouras

Athanasios Skouras is an external collaborator of the Department of Nursing of the Hellenic Mediterranean University. He received his Ph.D. in Pharmacy from the University of Patras in 2017. His expertise lies in Pharmaceutical Technology and more specifically in drug delivery systems. His research interests mainly include the formulation of novel theranostic systems utilizing lipids and the synthesis and applications of biodegradable nanoparticles.

Preface

Two-dimensional structured materials and metallic NPs exhibit extraordinarily distinctive optical, thermal, electronic, and mechanical properties. In addition, thanks to their biocompatibility, low-toxicity, and antimicrobial property, they have been extensively applied in several biomedical applications such as biosensing, cell labeling, tissue engineering, drug and gene delivery systems, and mesenchymal stem cells differentiation. Motivated by these achievements, we demonstrate the present Special Issue, which focuses on the use of 2D materials and other nanoparticles in several biomedical applications. It contains ten original research contributions (two theoretical studies, seven research articles and one review article) providing a broad coverage of recent research progress and insights in the field of nanotechnology induced biomedicine, including studies on the development of bionanocomposites, biosensing, drug and gene delivery systems, as well as on the evaluation of the antimicrobial property of graphene derivatives and metallic nanoparticles. This themed collection is addressed to a broad research audience of chemists, biologists, pharmacologists, materials scientists, and researchers in the field of nanomedicine. We really appreciate the great efforts of all involved authors to prepare and submit their excellent manuscripts.

Minas M. Stylianakis and Athanasios Skouras
Editors

Article

Nanoindentation of Graphene/Phospholipid Nanocomposite: A Molecular Dynamics Study

Vladislav V. Shunaev [1] and Olga E. Glukhova [1,2,*]

1. Department of Physics, Saratov State University, 410012 Saratov, Russia; shunaevvv@sgu.ru
2. Institute for Bionic Technologies and Engineering, I.M. Sechenov First Moscow State Medical University (Sechenov University), 119991 Moscow, Russia
* Correspondence: glukhovaoe@sgu.ru; Tel.: +7-8452-514562

Abstract: Graphene and phospholipids are widely used in biosensing and drug delivery. This paper studies the mechanical and electronic properties of a composite based on two graphene flakes and dipalmitoylphosphatidylcholine (DPPC) phospholipid molecules located between them via combination of various mathematical modeling methods. Molecular dynamics simulation showed that an adhesion between bilayer graphene and DPCC increases during nanoindentation of the composite by a carbon nanotube (CNT). Herewith, the DPPC molecule located under a nanotip takes the form of graphene and is not destroyed. By the Mulliken procedure, it was shown that the phospholipid molecules act as a "buffer" of charge between two graphene sheets and CNT. The highest values of electron transfer in the graphene/DPPC system were observed at the lower indentation point, when the deflection reached its maximum value.

Keywords: graphene; phospholipids; molecular dynamics; nanoindentation; local stress; electron transfer

1. Introduction

Graphene is a two-dimensional allotropic modification of carbon with a thickness of one atom [1]. High biocompatibility [2], unique adsorption properties [3,4], and large surface area of graphene allow it to form a compatible interface with phospholipid molecules [5–8]. Both graphene and phospholipids separately are widely used in biosensing [9–13] and drug delivery [14–16]. It is predicted that the synergistic effect of graphene and phospholipids could be used in various biomedical devices [17–21].

Simulation of graphene-based composites' nanoindentation by molecular dynamic methods allows to study local characteristics of the considered objects, as it makes it possible to place a nanotip strictly above the surface of interest. Huang et al. MD research showed that the strength and hardness of bio-inspired nanocomposites decreased with the increasing length of graphene layers, while increasing the length of the graphene layer avoids the generation of dislocations at the edge of the graphene sheet [22]. Nanoindentation of Cu/Gr layered nanopillars allowed to find the dependence of the composite mechanical properties on the specified boundary conditions and anomalous extrinsic size effect—the weakening influence caused by the dislocations at the edges of graphene was compensated by hardening of graphene in the middle of the sheet [23]. Peng et al. demonstrated that the strength of a copper substrate during indentation dramatically depended on the number of graphene layers on its surface [24]. Simulation of lipid bilayer and graphene under indentation loads has shown that a graphene coating could effectively maintain the structural and physiological stability of a bio-nanohybrid [25]. Modeling of nanoindentation showed the strengthening effect of graphene coverage on a nickel substrate [26].

The object of this study is a composite based on two graphene flakes and dipalmitoylphosphatidylcholine (DPPC) phospholipid molecules located between them. Previously, the authors found the configurations of such composites that provide optimal current–voltage characteristics and electron transfer values [27]. The aim of this work is to simulate nanoindentation of the bilayer graphene/DPPC composite with subsequent evaluation of the mechanical and electronic properties during deflection.

2. Results
2.1. Atomistic Model

At the initial stage, two DPPC molecules containing 1 phosphorus atom, 1 nitrogen atom, 8 oxygen atoms, 40 carbon atoms, and 80 hydrogen atoms were placed between two graphene monolayer flakes with dimensions $L_x \times L_y$ = 25.5 Å × 35.51 Å. This composite structure acted as the supercell with translation vectors L_x, L_y (Figure 1a). This supercell contained 878 atoms, 618 of which belonged to graphene, and 260 to the DPPC molecules. The geometric center of each DPPC molecule was located at a distance of 6.23 Å from each graphene sheet. The atomic structure of the designed supercell and the values of the translation vectors were optimized by the self-consistent charge density functional tight-binding (SCC DFTB) 2 method.

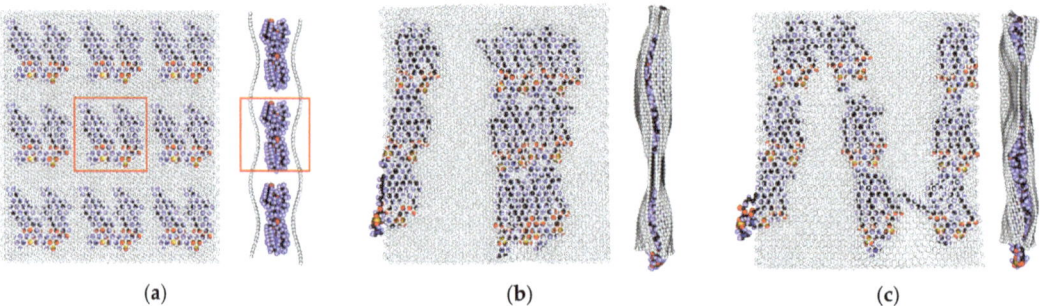

Figure 1. Atomic structure of the graphene/dipalmitoylphosphatidylcholine (DPPC) composite: (**a**) before the optimization; (**b**) type I after the optimization; and (**c**) type II after the optimization. The supercell obtained by the self-consistent charge density functional tight-binding (SCC DFTB) 2 method is highlighted by the red box. Graphene atoms are grey, phosphorous—yellow, nitrogen—blue, oxygen—red, carbon in phospholipid—black, and hydrogen—blue.

The fragment of a graphene/DPPC composite film containing 7902 atoms was built from nine optimized supercells. The fragment sizes were 90.46 Å × 80.56 Å, and the initial structure is shown in Figure 1a. As the search for the ground state of such a polyatomic fragment by the SCC DFTB 2 quantum mechanical method is practically impossible, the AMBER empirical method was applied. It should be noted that the fragment is not a supercell, but a finite structure. From the result of 20 numerical experiments of the graphene/DPPC composite optimization, two types of object's topology were identified. In the type I topology, 12 DPPC molecules formed a disordered bundle (Figure 1b). If an additional number of DPPC molecules are added to the composite, double and triple layers of lipids can be formed in this structure [5]. In the type II topology, the phospholipids were arranged in a shape vaguely resembling a spiral; part of one DPPC even left the pores between the graphene sheets (Figure 1c). During optimization, the total energy of the system decreased from 17.54, to 3.967 Mcal/mol and the adhesion energy between graphene flakes and DPPC molecules dropped from −896.115 to −4124.72 kcal/mol. Because, in this structure, two isolated DPPC molecules located in the center of the composite were clearly distinguished, it was chosen as the object of nanotip indentation.

2.2. Nanoindentation

At the next stage, the carbon nanotube (CNT) (16,0) with a closed edge was placed at a distance of 4.1 Å from the surface of the upper graphene sheet (Figure 2a). The length of the CNT was 50.62 Å. To perform the process of the composite material indentation, the nanotip approached the composite surface with a step of 0.5 Å. At each step, the process of relaxation scanning was started, the values of local stresses on the atoms were determined, and the adhesion energy between bilayer graphene and phospholipids was calculated. Based on the analysis of the above-mentioned characteristics, the so-called "key points" of indentation were identified (Table 1).

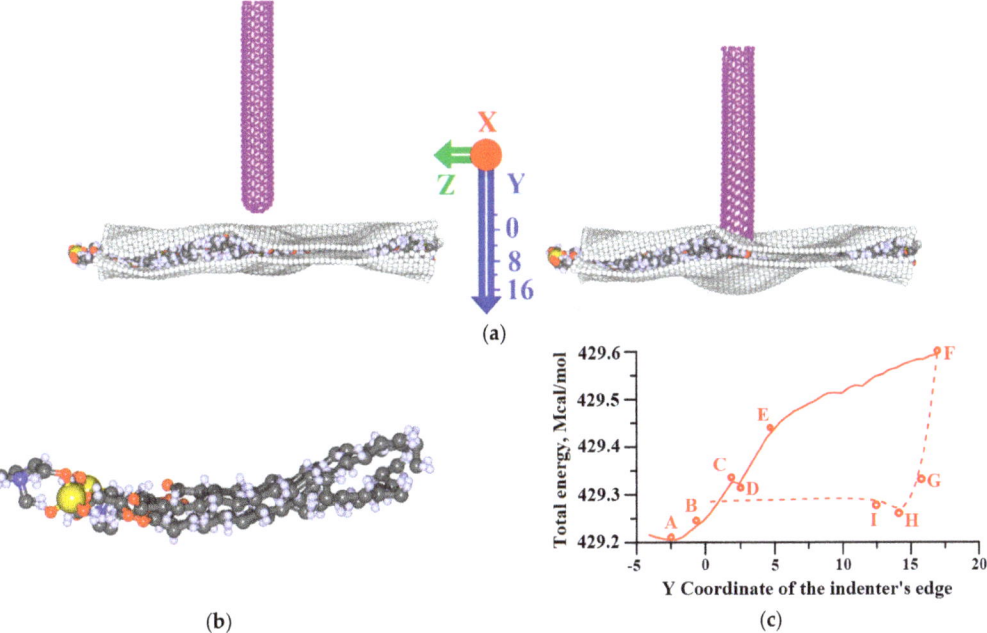

Figure 2. Nanoindentation of the composite graphene/DPPC by the carbon nanotube (CNT) with a closed edge: (**a**) initial (left) and last (right) point of indentation F; (**b**) the atomic structure of the central DPPC molecule at point F; and (**c**) the dependence of the system's total energy on the CNT shift. The solid line corresponds to the forward stroke (FS) and the dotted line corresponds to the reverse. At the initial point, the edge of the CNT had the coordinate Y = 0, and the atoms of the upper graphene layer had the coordinate Y = −4.1 Å.

Table 1. Key points of the nanoindentation with corresponding Y coordinates; the values of the adhesion energy between bilayer graphene and dipalmitoylphosphatidylcholine (DPPC) molecules and maximum local stresses (MLSs) in these points. The abbreviation FS corresponds to forward stroke and RS corresponds to reverse stroke.

Key Point	Y Coordinate of the Indenter's Edge, Å	Energy of Adhesion between Bilayer Graphene and Phospholipid, kcal/mol	MLS, GPa
A	−2.6 (FS)	−4125.1	0.2
B	−0.6 (FS)	−4124.03	0.47
C	1.9 (FS)	−4125.85	1.53
D	2.4 (FS)	−4127.13	1.71
E	4.4 (FS)	−4114.53	2.25
F	17.4 (FS)	−4116.33	2.53
G	15.4 (RS)	−4166.4	0.89
H	13.4 (RS)	−4172.05	0.97
I	12.4 (RS)	−4172.45	0.97

The dependence of the system's total energy on the CNT shift is shown in Figure 2c. A local minimum was observed at point A, indicating the most energetically favorable location between the CNT and the composite owing to the van der Waals (vdW) interaction. Further, the energy grew steadily until it reached point C. Between points C and D, there was a slight drop in the total energy of the system. In this region, the atomic structure of the DPPC molecule was rearranged and it started to take the form of graphene (Figure 2b). Note that, during the forward stroke, the adhesion energy between graphene and DPPC molecules reached its maximum value precisely at point D (Table 1). The position of CNTs at the lower indentation point F is shown in Figure 2a on the right. As the CNT left the trough of the composite (reverse stroke), the energy decreased greatly (from F to H). At this interval, the CNT tip stopped to pressurize the composite and began to relax. The point H corresponds to the moment of the strongest vdW interaction between objects during reverse stroke, so the local minimum of the total energy at this point is observed. Then, the vdW interaction became weaker and, from a certain moment (point I), the total energy practically did not change. Note that the trough created by the nanotip in the composite remained even after the CNT returned to its initial position. Herewith, the remaining 16 DPPC molecules did not leave the space between the graphene sheets (Figure 2a). The value of the adhesion energy between graphene and DPPC molecules at point I and further remained at the maximum for the entire time. Thus, despite the fact that no chemical bonds between graphene and DPPC molecules were formed during indentation, the bonds between graphene sheets and DPPC molecules were significantly strengthened.

2.3. Analysis of Graphene Sheets Strength during Nanoindentation

To assess the strength of graphene sheets during deflection, we calculated the map of the local stresses' distribution by atoms at each step (Figure 3a–h). As the maximum values of local stresses (MLSs) were found in the central atoms of the sheets (under nanotip), we presented maps only for these regions. Figure 3i shows the graph of the MLS on graphene atoms dependence on the indentation step. From the beginning of indentation to point B, the MLSs were observed in the upper sheet of graphene and varied from 0 to 0.47 GPa. Such minor changes were caused by the fact that the CNT has not yet reached the surface of the upper graphene sheet. The dependence on the BE section was almost linear, which indicated the elastic nature of the deformation. In this segment, the central phospholipid molecule under the CNT has taken the form of the curved graphene sheet. At the same time, there was significant increase in the MLS from 0.46 to 2.25 GPa. Therefore, at the BE segment, the graphene sheets started to provide a strengthening effect on the DPPC molecule, not allowing it to destruct under the influence of the nanotip. After this point, the adhesion between graphene and DPPC increased (see Section 2.2) and the MLS did not change much; on the EF segment, the MLS increased from 2.25 to 2.53 GPA. During the reverse stroke, a sharp drop in the MLS from 2.53 to 0.86 GPA was observed between F and G. At this interval, the composite started to relax because pressure from the tip became weaker (see Section 2.2). Starting from point G, the MLSs were observed in the lower graphene sheet and, starting from point H, the MLS values stopped changing because the energy of the vdW interaction between objects reached a minimum.

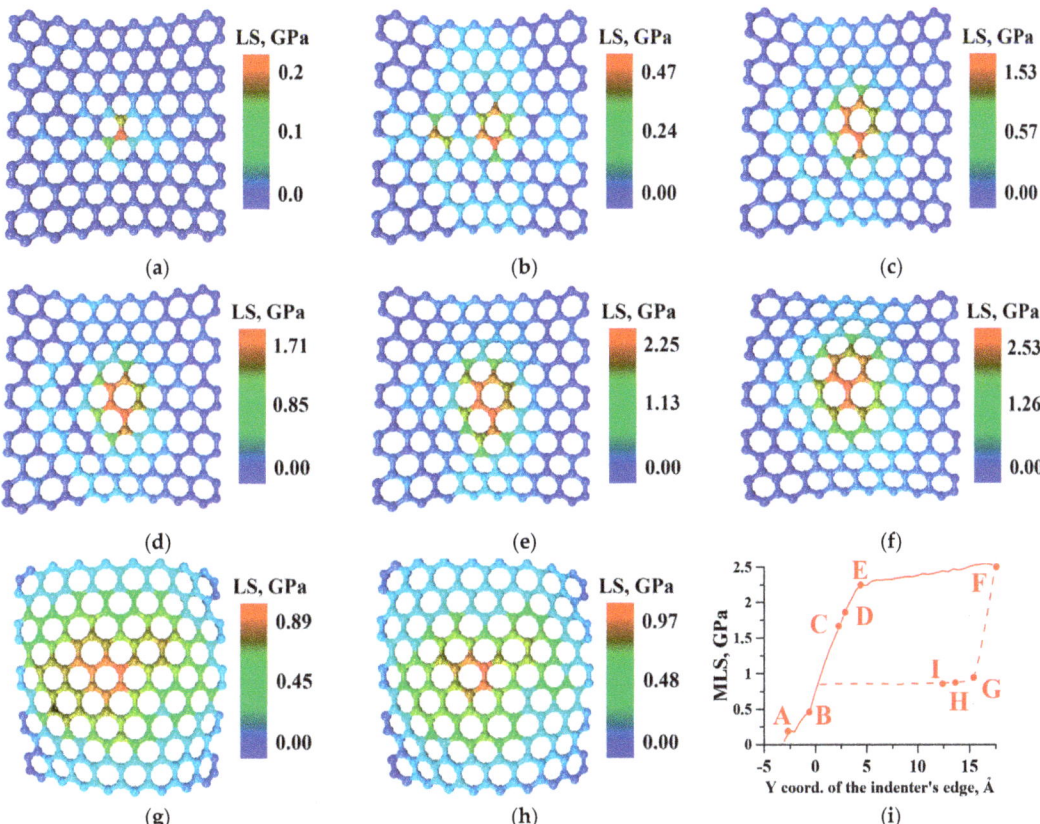

Figure 3. The map of the local stress (LS) in the central area of the graphene sheets at the key points of indentation: (**a**) point A; (**b**) point B (upper layer); (**c**) point C (upper layer); (**d**) point D (upper layer); (**e**) point E (upper layer); (**f**) point F (upper layer); (**g**) point G (lower layer); (**h**) point H (lower layer); and (**i**) the dependence of the maximum local stress (MLS) in graphene sheet atoms on the indentation step (the solid line corresponds to the forward stroke and the dotted line corresponds to the reverse).

2.4. Analysis of Electron Transfer in the CNT/Graphene/DPPC System during Nanoindentation

As the calculation of atoms' Mulliken charges by the SCC DFTB 2 method is resource consuming, the central part was cut out from the graphene/DPPC composite. This fragment containing two isolated DPPC molecules directly contacted the CNT and was responsible for charge transfer in the system (Figure 4a).

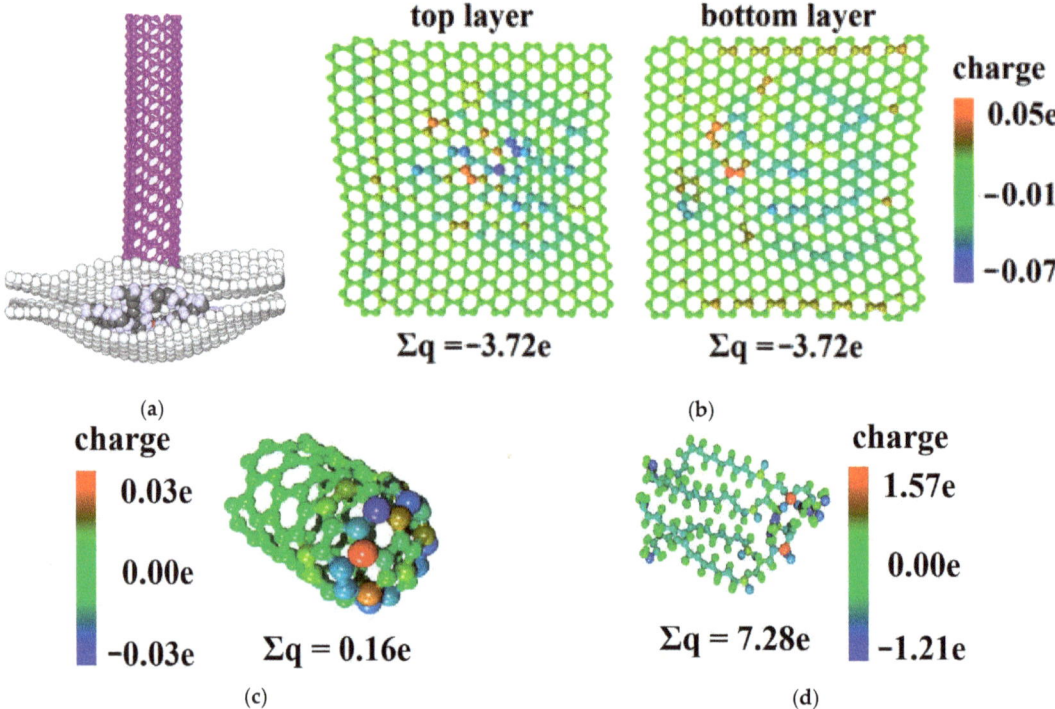

Figure 4. The analysis of the electron transfer in the CNT/graphene/DPPC system during nanoindentation (**a**) The fragment of the atomic structure graphene/DPPC with CNT at point F; (**b**) the distribution of Mulliken charges on atoms in the point for the upper (left) and the lower (right) graphene sheets; (**c**) for CNT; and (**d**) for DPPC.

The charge distributions between the atoms at the initial moment of time, at the lower point of the forward stroke F, and at the end of the reverse stroke I are shown in Table 2 At the initial moment of time, the CNT was electrically neutral. In the composite, the DPPC molecules acted as a donor and transferred the charge of 6.09 e to graphene sheets; the largest part of the charge was lost by two P-atoms (1.61 and 1.65 e), which is consistent with the results obtained earlier [26]. The charge between the graphene sheets was distributed unevenly because of the orientation of the DPPC molecules after optimization—the phosphorus molecules that tend to give charge and act as donors were located closer to the upper graphene sheet. During the indentation, the CNT gradually transferred the charge to the composite; at the lower point, its value was 0.16 e (Figure 4c). It can be seen that the phospholipid transferred even more charge to graphene sheets than in the initial state (7.28 e) (Figure 4d). The charge transferred from DPPC was evenly distributed between the graphene sheets (Figure 4b). This was caused by the fact that the DPPC molecules have taken the form of graphene sheets, as shown in Section 2.2, and the P atoms were located at the same distance from the graphene sheets. As the CNT is removed from the composite, it recovered the transferred charge and eventually became almost electroneutral again. Note that, during the reverse stroke, the main part of the charge came from the upper sheet of graphene that was directly in contact with the CNT. Thus, the DPPC molecules stopped to act as a "buffer" of charge between the two graphene sheets. The highest values of electron transfer in the graphene/DPPC system were observed at the lower indentation point, when the deflection reached its maximum value. Based on the conclusions of [26], it can be concluded that the deflection strain would significantly affect the current–voltage characteristics of the considered composite.

Table 2. Distribution of Mulliken charge in the system CNT/graphene/DPCC at different moments of nanoindentation. CNT, carbon nanotube.

	CNT	Upper Layer of Graphene Sheet	DPPC	Lower Layer of Graphene Sheet
Initial moment of time	0.00 e	−3.45 e	6.09 e	−2.63 e
Point F	0.16 e	−3.72 e	7.28 e	−3.72 e
Point I	−0.01 e	−3.28 e	6.87 e	−3.58 e

3. Methods

The search for the ground state of the graphene/DPPC composite film, as well as the study of its atomic structure changes during deflection by a nanotip, was performed by the AMBER empirical method [28] implemented in the Hyperchem software package [29]. Optimization was performed by the conjugate gradient Fletcher–Reeves method, and the root mean square (RMS) gradient was 0.1 kcal/(A·mol).

The adhesion energy between the phospholipids and graphene sheets at various steps was calculated by the following formula (Equation (1)):

$$E_{ADH} = E_{TOT} - E_{DPPC} - E_{UP} - E_{LOW} \quad (1)$$

where E_{TOT} is the energy of the graphene/DPPC system at this point, and E_{DPPC}, E_{UP}, and E_{LOW} are the energies of isolated DPPC, upper, and lower graphene sheets, respectively.

To estimate the strength of graphene sheets during deflection, the previously developed original method for calculating local stresses of the atomic grid was used [30]. According to this method, the stress σ_i on each atom is calculated by the formula $\sigma_i = |w_i - w_0|$, where w_0 is the energy volume density of the graphene atom before indentation, and w_i is the energy volume density of the graphene atom under external influence. The energy volume density of the atom was calculated by the formula $w_i = \frac{E_i}{V_i}$, where E_i is the energy of the atom calculated within the AMBER force field, and $V_i = \frac{4\pi r^3}{3}$ is the volume of the carbon atom (r = 1.7 Å).

The study of changes in the electronic structure during indentation and electronic transfer between phospholipid molecules and graphene was performed by Mulliken population analysis [31]. According to this method, the charge on an atom is calculated as the difference between the atomic number Z_A and GAP_A—the sum of the gross orbital product over all orbitals belonging to atom A: $Z = Z_A - GAP_A$. The charges were calculated by the quantum mechanical self-consistent charge density functional tight-binding (SCC DFTB) method [32] in the dftb+ software package [33] in the 3ob-3-1 basis.

4. Conclusions

The nanoindentation of the composite on the base of bilayer graphene and 16 DPPC phospholipid molecules was simulated by the molecular dynamics method. It was noted that, during indentation, the adhesion between graphene flakes and DPPC molecules increased and reached a maximum at the end of the reverse stroke. The maximum values of local stresses (in the region of 2.53 GPA) were observed on the upper graphene layer at the lower indentation point. At this moment, the DPPC molecule located under the nanotip took the form of curved graphene, while chemical bonds of the phospholipid molecules were not destroyed. The pressure of the CNT tip led to the growth of adhesion energy between graphene sheets and DPPC molecules. It is known that, getting into blood vessels, drug carriers sense abnormally high shear stresses [34,35], so the discovered effect of "phospholipid strengthening" by the graphene sheets' coating can be used in the field of drug delivery. It was found that the electron transfer between CNT and graphene/DPPC composites increased during indentation and reached 0.16 e. At this moment, the phospholipid molecule stopped to act as a "buffer" of charge between the two graphene sheets. The observed phenomenon of electronic transfer between graphene and phospholipid can be applied in biosensorics.

Author Contributions: Conceptualization, O.E.G.; methodology, O.E.G.; software, V.V.S.; validation, O.E.G. and V.V.S.; formal analysis, O.E.G.; investigation, V.V.S. and O.E.G.; resources, O.E.G.; data curation, V.V.S.; writing—original draft preparation, V.V.S. and O.E.G.; writing—review and editing, V.V.S.; visualization, V.V.S.; supervision, O.E.G.; project administration, O.E.G.; funding acquisition, O.E.G. All authors have read and agreed to the published version of the manuscript.

Funding: This research was funded by the Ministry of Science and Higher Education of the Russian Federation (project no. FSRR-2020-0004).

Institutional Review Board Statement: Not applicable.

Informed Consent Statement: Not applicable.

Data Availability Statement: The data presented in this study is available in this article.

Conflicts of Interest: The authors declare no conflict of interest.

References

1. Geim, A.K.; Novoselov, K.S. The rise of graphene. *Nat. Mater.* **2007**, *6*, 183–191. [CrossRef] [PubMed]
2. Pinto, A.M.; Goncalves, I.C.; Magalhaes, F.D. The rise of graphene. *Colloid. Surface. B* **2013**, *11*, 188–202. [CrossRef] [PubMed]
3. Stergiou, A.; Cantón-Vitoria, R.; Psarrou, M.N.; Economopoulos, S.P.; Tagmatarchis, N. Functionalized graphene and targeted applications—Highlighting the road from chemistry to applications. *Prog. Mater. Sci.* **2020**, *114*, 100683.
4. Nandanapalli, K.R.; Mudusu, D.; Lee, S. Functionalization of graphene layers and advancements in device applications. *Carbon* **2019**, *152*, 954–985. [CrossRef]
5. Rivela, T.; Yesylevskyy, S.O.; Ramseyer, C. Structures of single, double and triple layers of lipids adsorbed on graphene: Insights from all-atom molecular dynamics simulations. *Carbon* **2017**, *118*, 358–369. [CrossRef]
6. Lima, L.M.C.; Fu, W.; Jiang, L.; Kros, A.; Schneider, G.F. Graphene-stabilized lipid monolayer heterostructures: A novel biomembrane superstructure. *Nanoscale* **2016**, *8*, 18646–18653.
7. Li, W.; Moon, S.; Wojcik, M.; Xu, K. Direct Optical Visualization of Graphene and Its Nanoscale Defects on Transparent Substrates. *Nano Lett.* **2016**, *16*, 5022–5026. [CrossRef]
8. Okamoto, Y.; Tsuzuki, K.; Iwasa, S.; Ishikawa, R.; Sandhu, A.; Tero, R. Fabrication of Supported Lipid Bilayer on Graphene Oxide. *J. Phys. Conf. Ser.* **2012**, *352*, 012017. [CrossRef]
9. Mitsakakis, K.; Sekula-Neuner, S.; Lenhert, S.; Fuchs, H.; Gizeli, E. Convergence of Dip-Pen Nanolithography and Acoustic Biosensors towards a Rapid-Analysis Multi-Sample Microsystem. *Analyst* **2012**, *137*, 3076–3082. [CrossRef]
10. Bog, U.; Laue, T.; Grossmann, T.; Beck, T.; Wienhold, T.; Richter, B.; Hirtz, M.; Fuchs, H.; Kalt, H.; Mappes, T. On-Chip Microlasers for Biomolecular Detection via Highly Localized Deposition of a Multifunctional Phospholipid Ink. *Lab Chip* **2013**, *13*, 2701–2707. [CrossRef]
11. Bog, U.; Brinkmann, F.; Wondimu, S.F.; Wienhold, T.; Kraemmer, S.; Koos, C.; Kalt, H.; Hirtz, M.; Fuchs, H.; Koeber, S.; et al. Densely Packed Microgoblet Laser Pairs for Cross-Referenced Biomolecular Detection. *Adv. Sci.* **2015**, *2*, 1500066. [CrossRef] [PubMed]
12. Lahcen, A.A.; Rauf, S.; Beduk, T.; Durmus, C.; Aljedaibi, A.; Timur, S.; Alshareef, H.N.; Amine, A.; Wolfbeis, O.S.; Salama, K.N.; et al. Electrochemical sensors and biosensors using laser-derived graphene: A comprehensive review. *Biosens. Bioelectron.* **2020**, *168*, 112565. [CrossRef] [PubMed]
13. Walther, B.K.; Dinu, C.Z.; Guldi, D.M.; Sergeyev, V.G.; Creager, S.E.; Cooke, J.P.; Guiseppi-Elie, A. Nanobiosensing with graphene and carbon quantum dots: Recent advances. *Mater. Today* **2020**, *39*, 23–46. [CrossRef]
14. Chauhan, G.; Shaik, A.A.; Kulkarni, N.S.; Gupta, V. The preparation of lipid-based drug delivery system using melt extrusion. *Drug Discov. Today* **2020**, *25*, 1930–1946.
15. Singh, R.P.; Gangadharappa, H.V.; Mruthunjaya, K. Phospholipids: Unique carriers for drug delivery systems. *J. Drug Deliv. Sci. Technol.* **2017**, *39*, 166–179. [CrossRef]
16. Song, S.; Shen, H.; Wang, Y.; Chu, X.; Xie, J.; Zhou, N.; Shen, J. Biomedical application of graphene: From drug delivery, tumor therapy, to theranostics. *Colloid. Surface. B* **2020**, *185*, 110596. [CrossRef]
17. Willems, N.; Urtizberea, A.; Verre, A.F.; Iliut, M.; Lelimousin, M.; Hirtz, M.; Vijayaraghavan, A.; Sansom, M.S.P. Biomimetic Phospholipid Membrane Organization on Graphene and Graphene Oxide Surfaces: A Molecular Dynamics Simulation Study. *ACS Nano* **2017**, *11*, 1613–1625.
18. Monasterio, B.G.; Alonso, B.; Sot, J.; García-Arribas, A.B.; Gil-Cartón, D.; Valle, M.; Zurutuza, A.; Goñi, F.M. Coating graphene oxide with lipid bilayers greatly decreases its hemolytic properties. *Langmuir* **2017**, *33*, 8181–8191. [CrossRef]

19. Durso, M.; Borrachero-Conejo, A.I.; Bettini, C.; Treossi, E.; Scidà, A.; Saracino, E.; Gazzano, M.; Christian, M.; Morandi, V.; Tuci, G.; et al. Biomimetic graphene for enhanced interaction with the external membrane of astrocytes. *J. Mater. Chem. B* **2018**, *6*, 5335–5342. [CrossRef]
20. Hai, L.; He, D.; He, X.; Wang, K.; Yang, X.; Liu, J.; Cheng, H.; Huang, X.; Shangguan, J. Facile fabrication of a resveratrol loaded phospholipid@reduced graphene oxide nanoassembly for targeted and near-infrared laser-triggered chemo/photothermal synergistic therapy of cancer in vivo. *J. Mater. Chem. B* **2017**, *5*, 5783–5792. [CrossRef]
21. Kuo, C.J.; Chiang, H.C.; Tseng, C.A.; Chang, C.; Ulaganathan, R.K.; Ling, T.-T.; Chang, Y.-J.; Chen, C.-C.; Chen, Y.-R.; Chen, Y.-T. Lipid-modified graphene-transistor biosensor for monitoring amyloid- aggregation. *ACS Appl. Mater. Interfaces* **2018**, *10*, 12311–12316. [CrossRef]
22. Huang, Y.; Yanga, Z.; Lu, Z. Nanoindentation of bio-inspired graphene/nickel nanocomposites: A molecular dynamics simulation. *Comput. Mater. Sci.* **2021**, *186*, 109969. [CrossRef]
23. Shuang, F.; Aifantis, K.E. Dislocation-graphene interactions in Cu/graphene composites and the effect of boundary conditions: A molecular dynamics study. *Carbon* **2021**, *172*, 50–70. [CrossRef]
24. Peng, W.; Sun, K.; Zhang, M.; Shi, J.; Chen, J. Effects of graphene coating on the plastic deformation of single crystal copper nano-cuboid under different nanoindentation modes. *Mater. Chem. Phys.* **2019**, *225*, 1–7. [CrossRef]
25. Song, Z.; Wang, Y.; Xu, Z. Mechanical responses of the bio-nano interface: A molecular dynamics study of graphene-coated lipid membrane. *Theor. Appl. Mech. Lett.* **2015**, *5*, 231–235. [CrossRef]
26. Yan, Y.; Zhou, S.; Liu, S. Atomistic simulation on nanomechanical response of indented graphene/nickel system. *Comput. Mater. Sci.* **2017**, *130*, 16–20.
27. Slepchenkov, M.M.; Glukhova, O.E. Improving the Sensory Properties of Layered Phospholipid-Graphene Films Due to the Curvature of Graphene Layers. *Polymers* **2020**, *12*, 1710. [CrossRef]
28. Cornell, W.D.; Cieplak, P.; Bayly, C.I. A Second Generation Force Field for the Simulation of Proteins, Nucleic Acids, and Organic Molecules. *J. Am. Chem. Soc.* **1995**, *117*, 5179–5197. [CrossRef]
29. Chemistry Software, HyperChem, Molecular Modeling. Available online: http://www.hyper.com/ (accessed on 1 December 2020).
30. Slepchenkov, M.M.; Glukhova, O.E. Influence of the curvature of deformed graphene nanoribbons on their electronic and adsorptive properties: Theoretical investigation based on the analysis of the local stress field for an atomic grid. *Nanoscale* **2012**, *11*, 3335–3344.
31. Mulliken, R.S. Electronic Population Analysis on LCAO-MO Molecular Wave Functions. *J. Chem. Phys.* **1995**, *23*, 1833–1840.
32. Elstner, M.; Porezag, D.; Jungnickel, G.; Elsner, J.; Haugk, M.; Frauenheim, T.; Suhai, S.; Seifert, G. Self-consistent-charge density-functional tight-binding method for simulations of complex materials properties. *Phys. Rev. B* **1998**, *58*, 7260–7268. [CrossRef]
33. DFTB+ Density Functional Based Tight Binding. Available online: https://dftbplus.org (accessed on 1 December 2020).
34. Xia, Y.; Shi, C.-Y.; Xiong, W.; Hou, X.-L.; Fang, J.G.; Wang, W. Shear Stress-sensitive Carriers for Localized Drug Delivery. *Curr. Pharm. Des.* **2016**, *22*, 5855–5867. [PubMed]
35. Godoy-Gallardo, M.; Ek, P.K.; Jansman, M.M.T.; Wohl, B.M.; Hosta-Rigau, L. Interaction between drug delivery vehicles and cells under the effect of shear stress. *Biomicrofluidics* **2015**, *9*, 052605. [CrossRef] [PubMed]

Article

Theoretical Study on the Aggregation and Adsorption Behaviors of Anticancer Drug Molecules on Graphene/Graphene Oxide Surface

Pengyu Gong, Yi Zhou, Hui Li, Jie Zhang, Yuying Wu *, Peiru Zheng * and Yanyan Jiang *

Key Laboratory for Liquid-Solid Structural Evolution and Processing of Materials, Ministry of Education, Shandong University, Jinan 250061, China
* Correspondence: wuyuying@sdu.edu.cn (Y.W.); sduzpr3@outlook.com (P.Z.); yanyan.jiang@sdu.edu.cn (Y.J.)

Abstract: Graphene and its derivatives are frequently used in cancer therapy, and there has been widespread interest in improving the therapeutic efficiency of targeted drugs. In this paper, the geometrical structure and electronic effects of anastrozole(Anas), camptothecin(CPT), gefitinib (Gefi), and resveratrol (Res) on graphene and graphene oxide(GO) were investigated by density functional theory (DFT) calculations and molecular dynamics (MD) simulation. Meanwhile, we explored and compared the adsorption process between graphene/GO and four drug molecules, as well as the adsorption sites between carriers and payloads. In addition, we calculated the interaction forces between four drug molecules and graphene. We believe that this work will contribute to deepening the understanding of the loading behaviors of anticancer drugs onto nanomaterials and their interaction.

Keywords: DFT calculations; MD simulations; adsorption and aggregation; graphene; graphene oxide; anticancer drugs

Citation: Gong, P.; Zhou, Y.; Li, H.; Zhang, J.; Wu, Y.; Zheng, P.; Jiang, Y. Theoretical Study on the Aggregation and Adsorption Behaviors of Anticancer Drug Molecules on Graphene/Graphene Oxide Surface. *Molecules* 2022, 27, 6742. https://doi.org/10.3390/molecules27196742

Academic Editors: Minas M. Stylianakis and Athanasios Skouras

Received: 29 August 2022
Accepted: 6 October 2022
Published: 10 October 2022

Publisher's Note: MDPI stays neutral with regard to jurisdictional claims in published maps and institutional affiliations.

Copyright: © 2022 by the authors. Licensee MDPI, Basel, Switzerland. This article is an open access article distributed under the terms and conditions of the Creative Commons Attribution (CC BY) license (https://creativecommons.org/licenses/by/4.0/).

1. Introduction

Nanomaterials have broad application prospects in the biomedical field because of their unique characteristics. Drug delivery based on nanoparticles has been extensively studied to maximize the therapeutic efficacy of drugs [1]. Among the diverse nanomaterials that have been found, graphene and its derivatives have been demonstrated to provide efficient drug delivery and are considered as promising and ideal nanocarriers for drug delivery systems, and have been widely studied in the field of cancer treatment [2] due to their remarkable physical and chemical properties [2]. Graphene is a two-dimensional (2D) sheet of sp^2 hybrid carbon atoms; the carbon atoms are tightly packed in a (2D) honeycomb lattice, which exhibits excellent properties such as large surface area and good biocompatibility, as well as providing a defect-free plane [3]. These pivotal characteristics allow it to interact with drugs through non-covalent interactions such as π-π interaction. As a derivative of graphene, GO is also a promising drug delivery vehicle [4]. Apart from features similar to pristine graphene, abundant hydroxyl, epoxy, and carboxyl functional groups in GO enable it to have a higher adsorption capacity for drug molecules than pristine graphene [5].

There are many drugs that could be delivered by graphene and its oxide, such as camptothecin, a widely used anticancer drug [6,7]. Its main target in cells is the type I DNA topoisomerase, which can inhibit DNA synthesis through chain break, causing cell death during the S phase of the cell cycle, making it an effective inhibitor of leukemia cell growth [8–10]. Based on this mechanism, Liu et al. studied the inhibitory effect of CPT on the growth of prostate cancer cells, as it can selectively inhibit the androgen-responsive growth of prostate cancer cells [11]. Therefore, CPT is also a potential candidate drug for the treatment of prostate cancer. Furthermore, resveratrol is a phytoalexin extracted in many edible plants that may play a role in preventing inflammation, atherosclerosis, cancer, and so forth [12,13]. For example, Kueck et al. found that Res inhibits glucose metabolism in human ovarian cancer cells. Zhou et al. [14] proposed that Res can induce apoptosis of pancreatic cancer cells. These studies suggest that Res is an effective cancer drug. In addition, gefitinib, an oral epidermal growth factor receptor (EGFR) tyrosine kinase inhibitor, is the first approved targeting drug for the treatment of non-small cell lung cancer (NSCLC) [15]. It has also been widely studied as a prospective drug for other cancers besides NSCLC, Li et al. [16] studied the potential role of Gefi in the treatment of pancreatic cancer and found that it can inhibit the growth, invasion, and colony formation of pancreatic cancer cells/Kalykaki et al. evaluated the effects of Gefi on circulating tumor cells (CTCs) in patients with metastatic breast cancer (MBC) [17,18]. Last but not least, anastrozole is a third generation aromatase inhibitor. As a potent inhibitor of intratumoral estrogen [19], clinical trials showed that Anas reduced the risk of breast cancer in postmenopausal women by 53% [20]. In the treatment of advanced breast cancer, it has significant survival benefits and tolerability advantages compared to other treatment drugs [21]. Therefore, it plays an important role in the prevention and treatment of breast cancer [22].

In this paper, the adsorption behavior of these drugs on graphene and GO carriers was investigated in depth using density functional theory (DFT) and molecular dynamics (MD) simulation, aiming to find the most stable adsorption conformations of different drug molecules on graphene and GO, and to compare the adsorption performance of the same carrier for different drugs. We hope that the results of this study can provide significant value for further design and development of new nanomaterial drug delivery systems, which we believe will ultimately improve the efficacy of targeted drugs in cancer therapy.

2. Computational Methods

2.1. Quantum Chemistry Calculations

We used quantum chemistry calculations methods to investigate the energetics of graphene and GO, and the effect of adsorption on drugs. DFT calculation is a quantum mechanical approach to study electronic systems and is commonly used to calculate the bind band structure of solids in physics. This method has been used for graphene-related research [23] All the quantum chemistry calculations were carried out with the Atomistix ToolKit (ATK) package. Generalized gradient approximation (GGA) [24] with Perdew–Burke–Ernzerhof (PBE) parametrization [25,26] was used as the exchange-correlation functional. The basis set consists of the double numerical atomic orbitals augmented by polarization functions, which are comparable to Gaussian 6–31G**. Compared with other methods, this calculation method is more effective and can meet the accuracy requirements [27,28]. To avoid neighboring interaction, the distance between the neighboring molecules was larger than 15 Å. The real-space global cutoff radii were set as 3.7 Å. The convergence criterion on the energy and electron density was set to be 10^{-5} hartree. Geometry optimizations were performed with convergence criteria of 2×10^{-3} hartree/Å on the gradient, and 5×10^{-3} Å on the displacement.

The adsorption energy of CPT on to the studied nanosheets and GO is calculated using the relation:

$$E_a = E_{complex} - E_{nanosheet} - E_{drug} \tag{1}$$

where $E_{complex}$, $E_{carrier}$, and E_{drug} are the total energy of the complex, energy of the carrier (GRA or GO), and energy of the drug molecule (Res, Ana, Gefi, or CPT).

2.2. Molecular Dynamics Simulation

MD simulation is a method of simulating molecules in chemistry using classical Newtonian mechanics with computer simulations [29] to obtaining material properties. MD has been widely used in the calculation of the materials such as graphene. Due to problems such as speed and difficulty in calculating large systems, we chose classical molecular dynamics as our research method. The force-field parameters were taken from the CHARMM force-field. We used the SwissParam web server to obtain the force-field parameters of the drug molecules. All the simulations were carried out by using the GROMACS 2018 software package. The initial structure of graphene containing 480 carbon atoms was constructed with the Nanotube Modeler package. To create GO, we randomly decorate the graphene surface with hydroxyl and epoxy groups. The final oxygen to carbon (O/C) ratio of GO nanosheets is 1:8. As for the relaxation of drug molecules, a box with a size of 4 nm × 4 nm × 4 nm was firstly established, small molecules were randomly inserted into the box, and the steepest descent method was used to optimize the system to remove close contact and overlapping. Since both sides of the graphene can be used for drug binding, it was placed in the middle of the box and the drug molecules were allowed to be randomly distributed on both sides. In all systems, the center of the graphene sheet was set as the zero point. Each system performed 10 ns NVT relaxation at 298 K and 1 atm, followed by 10 ns NPT-relaxation. After that, 50 ns MD simulation was conducted at 298 K and 1 atm equilibrium, and the integration step was 2.0 fs. The Berendsen thermal bath method was employed to control the temperatures. The cutoff radius of non-bonding interaction was set as 1.4 nm. Trajectories were collected every 5000 steps for further analysis. Visual molecular dynamics (VMD) was used to observe the movement trajectory of the system.

3. Results and Discussion

3.1. Electrostatic Potential (ESP) of Drug Molecules

The reactivity and the interaction (especially for non-covalent interaction) of molecules can be determined by molecular surface electron density and electron activity, which is usually described by molecular electrostatic potential (ESP). In order to unveil the possible active sites in different drug molecules during drug adsorption, we have drawn electrostatic potential diagrams of different molecules, as shown in Figure 1. The red region represents positive electrostatic potential and shows electrophilicity, while the blue region represents negative electrostatic potential, which is more nucleophilic.

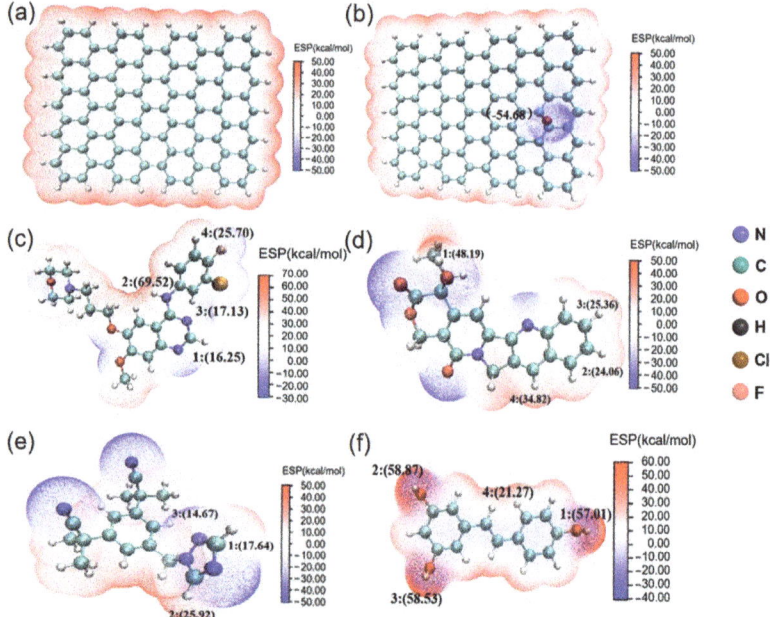

Figure 1. The electrostatic potential (ESP) distribution of (**a**) RGO; (**b**) GO; (**c**) Gefi; (**d**) CPT; (**e**) Anas; (**f**) Res.

As shown in Figure 1a, graphene has uniform electron density and abundant π electrons on its surface. According to previous studies, graphene is prone to π-π electron donor acceptor interactions and van der Waals (vdW) interactions due to its large ring plane structure [30]. The graphene oxide shown in Figure 1b is plotted in blue at the oxygen atom. GO has reduced π electron activity to some extent due to the presence of oxygen-containing functional groups, but it may form hydrogen bonds with other molecules. The oxygen-containing functional groups of GO possess higher chemical reactivity compared to graphene. The ESP of Gefi is shown in Figure 1c, with lower ESP at the oxygen atom position. The ESP distribution of CPT is the same as that of Gefi, as shown in Figure 1d, with lower ESP near the functional group, which is more nucleophilic compared to the position of the hydrogen atom. The nitrogen atom position is shown in Figure 1e. The nitrogen atom position is plotted in blue with lower ESP, as shown in Figure 1f, and the oxygen atom position is plotted using red, indicating that the point has higher ESP [31].

3.2. Simulation of Graphene Adsorption of Drug Molecules on Graphene

As for the adsorption energy between graphene and drug molecules, we mainly focus on the parallel configuration of drug molecules due to the abundant π electrons on the graphene, which would easily result in the adsorption of drug molecules through vdW interactions. The optimized structures of graphene after adsorbing four drug molecules and their adsorption energy are shown in Figure 2.

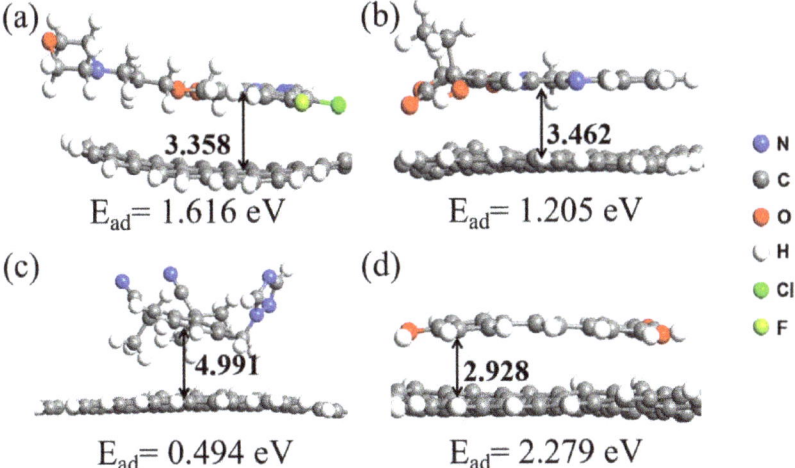

Figure 2. Optimized geometries of RGO and drug systems; bonds are in Å (the vertical distance refers to the distance between the centroid of benzene of different drugs to the carbon plane of RGO). (**a**) Graphene-Gefi; (**b**) Graphene-Camptothecin; (**c**) Graphene-Anas; (**d**) Graphene-Res.

The vertical distance is the distance from the center of mass of the aromatic ring of the drug molecule to the plane of the graphene. The vertical distances are between the graphene and the aromatic rings of Gefi, CPT, Anas, Res are 3.358 Å, 3.462 Å, 4.991 Å, and 2.928 Å. The distance between graphene and different drug molecules follows the order: Anas > CPT > Gefi > Res; and their adsorption energy is Anas < CPT < Gefi < Res from small to large, indicating that the greater distance, the weaker the adsorption capacity of the drug is [32–35].

We used MD simulations to study the effect of drugs adsorption on the graphene and GO. We employed the root mean square deviation (RMSD), density distribution, radial distribution function (RDT), hydrogen bond number (H- bond) and mean square displacement (MSD) to investigate the dynamics process of the adsorption of drug molecules on graphene.

Figure 3a shows the RMSD curves for different systems to investigate the equilibrium state of the simulated system. It can be seen that there are no large fluctuations in the RMSD curves of the system at the later stages of the simulation, indicating that it is sufficient to bring the system to equilibrium within the simulation time. As stated above, the main driving force for the adsorption of different drug molecules by graphene stems from the π-π interactions. The distribution of drug molecules on both sides of graphene was first investigated and the results are shown in Figure 3b. Their mass density shows that different drug molecules are effectively adsorbed on both side of the graphene sheet after the adsorption has reached equilibrium and that their distribution on both sides is symmetrical. Their two symmetry peaks are both at a distance of 0.38 nm, which is close to the vdW radius of the carbon atoms on the graphene sheet. In the range of distance less than 0.5 nm, all four kinds of drug molecules could appear on both sides of the graphene sheet. No drug molecule was observed beyond the range of 1 nm after the equilibrium of the system, because the mass densities of the drug molecules were all close to zero when the distances greater than 1 nm, which also indicates that their adsorption is relatively tight. Radial distribution functions (RDF) can be used to study the intermolecular interaction. Figure 3c shows the interaction between drug molecules and graphene in the simulated system. There are significant interactions between different drug molecules and graphene. Their peaks are 0.482 nm for Gefi, 0.461 nm for CPT, 0.644 nm for Anas, and 0.436 nm for Res. There are only vdW interactions between the four drug molecules and graphene, and the strength of the interaction between drug molecules and graphene decreases as the distance increases. Res has the strongest interaction with graphene, while Anas has the weakest interaction. The adsorption capacities of graphene to the four drug molecules follow the order Res > CPT > Gefi > Anas. To further investigate the interaction between drug molecules and graphene, we calculated the probabilistic profiles of the distribution of the angle between the aromatic ring for drug molecules and the graphene plane in molecular dynamics. The ability of drug molecules to absorb on the graphene is mainly determined by the superposition of π-π interactions. Effective interactions between aromatic rings are considered to occur when the angle $\alpha < 30°$. Figure 3d shows that all angles between the aromatic rings of four drug molecules and the graphene plane are small during the simulations. The most observed angles between Gefi, CPT, Anas, Res and the graphene were approximately 7°, 7°, 13°, and 8°, which indicates that their aromatic rings are almost parallel to the graphene surface. This also demonstrates that the adsorptions between drug molecules and graphene are stable and that π-π interactions are the main driving force for this adsorption. The environment will have a strong influence on the motion of the molecules in the system. Figure 3e shows the MSD results for different drug molecules adsorbing on graphene. Table 1 show the self-diffusion coefficients of the drug molecules in different systems. It can be seen that the diffusion coefficient of Anas is the largest, and the diffusion coefficients of CPT, Gefi, and Res are relatively similar, indicating that the diffusive motion of Anas is the most active among the four drugs. This result can be attributed to the weak binding ability of Anas and its low molecule weight (293.73), which facilitates its diffusion, whereas the other three drugs were strongly absorbed on the graphene surface and thus diffused more slowly [36,37].

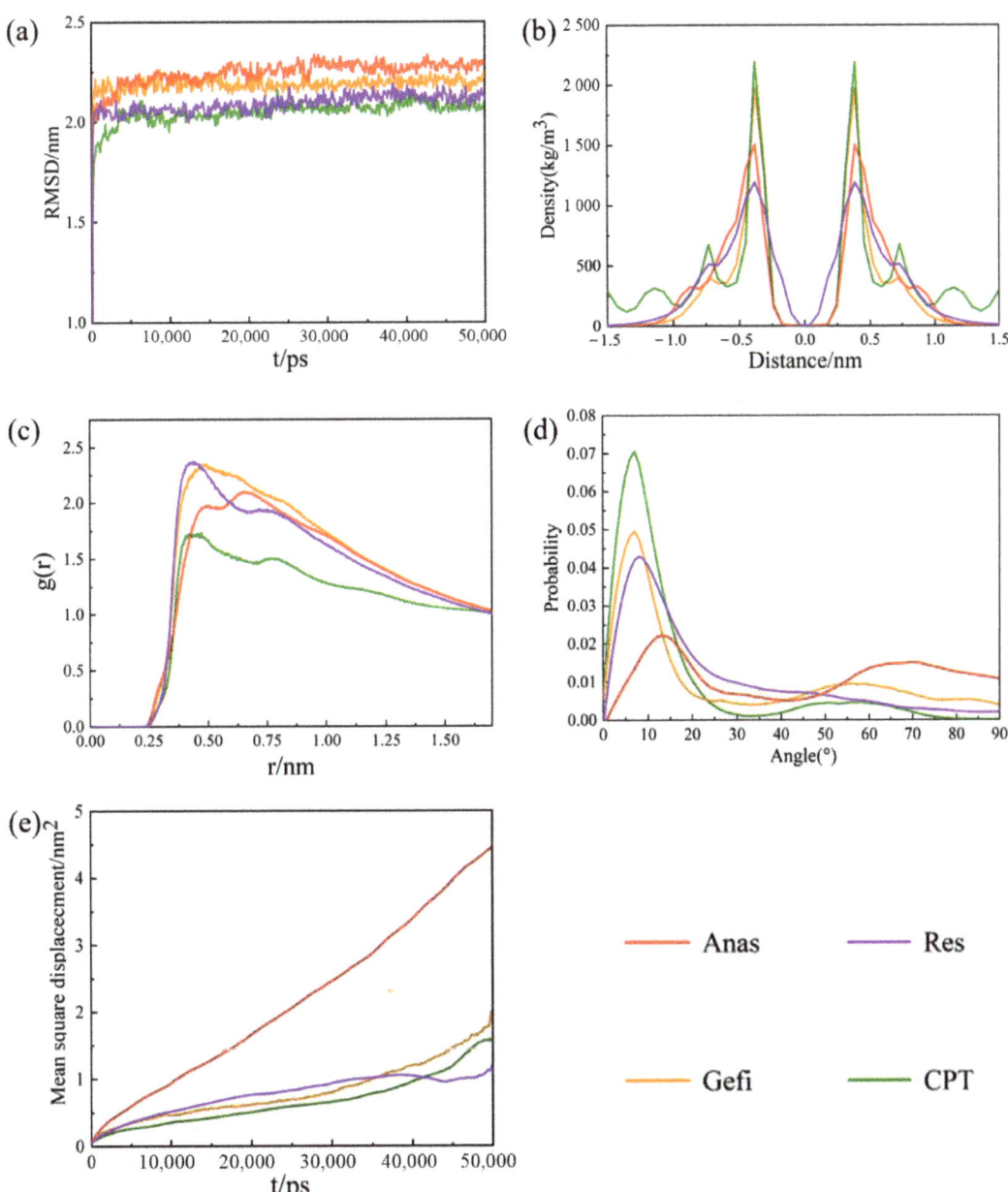

Figure 3. (**a**) The RMSD plots of the system with different drugs adsorbed on graphene as function of time. (**b**) Mass density profiles of different drugs. The center of graphene is set as distance = 0. (**c**) Radial distribution functions (RDF) of the different drug molecules with graphene. (**d**) The probability of the angle between the aromatic rings of the drug molecules and the graphene plane, (**e**) The MSD plots of the different drug molecules with graphene.

Table 1. Self-diffusion coefficients of different drugs adsorbed on graphene surface.

Drug Molecules	Gefi	CPT	Anas	Res
Diffusion coefficient ($\times 10^{-5}$ cm$^2\cdot$s^{-1})	0.004027	0.003376	0.0123	0.002793

3.3. Simulation of the Adsorption of Drug Molecules on GO

As for the adsorption behavior of drugs on GO, apart from the π-π interaction, drug molecules also generate hydrogen bonds with GO due to the functional groups, which promotes the adsorption of drug molecules on GO. According to Figure 1, we investigate the adsorption of different binding sites and named those structures as [GO—different drug names—X] (X = 1,2 ...). Several configurations were optimized and only two stable configurations for each kind of drug were selected to be further analyzed, which are illustrated in Figure 4.

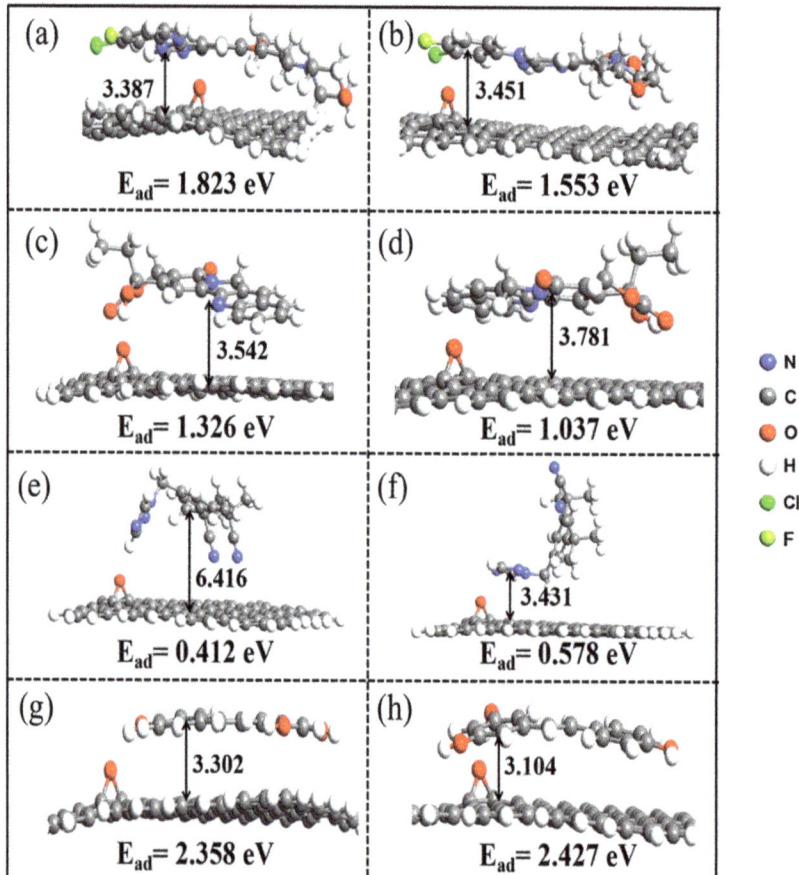

Figure 4. Optimized geometries of GO and drug systems; bonds are in Å (the vertical distance refers to the distance between the centroid of benzene of different drugs to the carbon plane of GO). (**a**) GO-Gefi-2; (**b**) GO-Gefi-4; (**c**) GO-Camptothecin-1; (**d**) GO-Camptothecin-4; (**e**) GO-Anas-1; (**f**) GO-Anas-2; (**g**) GO-Res-1; (**h**) GO-Res-2.

As shown in Figure 4a,b, the adsorption energies of stable configurations after Gefi adsorption on GO with active sites 2 and 4 of Gefi as binding sites are 1.823 eV and

1.553 eV, respectively, and the vertical distances are 3.387 Å and 3.451 Å between the aromatic ring of Gefi and GO, respectively. Thus, the adsorption of GO with Gefi is more stable when atom number 2 is used as the adsorption site, which is also consistent with the result of the electrostatic potential of molecule. Active sites 1 and 4 of the CPT were investigated as binding sites for adsorption onto GO, and their optimized result is shown as Figure 4c,d. The perpendicular distances between the aromatic ring of the CPT and the GO are 3.542 Å and 3.781 Å, respectively, and the adsorption energies are 1.326 eV and 1.037 eV, respectively; therefore, the configuration of the CPT molecules adsorbed on GO in Figure 4c can be considered as relatively stable. Figure 4e,f show the stable configurations obtained after adsorption on GO using active sites 1 and 2 of Anas as binding sites. The distances of the aromatic ring and GO are 6.416 Å and 3.431 Å, respectively. In Figure 4e, although the Anas molecule can form hydrogen bonds with GO, the vertical distance becomes larger, thus weakening the π-π interactions and leading to a reduction in the overall adsorption energy. Therefore, for Anas, the adsorption is more stable when binding at active site 2. As shown in Figure 1, all three active sites of Res have relatively large in electrostatic potential values and small differences; therefore, their adsorption with graphene is similar. Figure 4g,h show the result of the adsorption of active sites 1 and 2 of Res with GO. The configuration of the adsorption at active site 3 is not shown in the figure because it is consistent with that of active site 2. It can be seen that the vertical distance and adsorption energies of Res on GO are relatively close in both cases and, therefore, they are stable as binding sites for adsorption with GO in both cases. Overall, in the adsorption of four drug molecules onto GO, the final conformation is more stable when the binding site has higher electrostatic potential. According to the binding configuration and adsorption energy, the adsorption capacity of GO for the four drug molecules follows the order: Res > Gefi > CPT > Anas [38–42]. Comparing Figures 2 and 4, the adsorption capacity of GO for four drug molecules is generally better than that of graphene. The vertical distance between different drug molecules and GO increased slightly, which weakened the π-π interaction between them to some extent, but the formation of hydrogen bonds promoted the binding of the two and the superposition of the two effects finally promoted the adsorption of drug molecules on GO.

Figure 5a shows the RMSD result for the adsorption of four drug molecules on GO, from which it can be seen that all the systems reach the equilibrium state in a short time.

Previous studies have shown that not only π-π interactions, but also some hydrogen bonds occur during the adsorption of drug molecules on GO. To deeply understand the strength of the two kinds of interactions during the simulation, we investigated the average interaction energy between the four drug molecules and the GO sheet. As shown in Figure 5b. It can be seen that the vdW interaction accounts for the major part of the potential energy and is much greater than the Coulomb interaction, indicating that vdW interaction is the dominant force between GO and the drug molecules. This conclusion is also supported by the results of the radial distribution function of the simulated system [43].

Figure 5c shows the RDFs between different drug molecules and GO sheets. Four drug molecules had RDFs with GO in the range of less than 0.35 nm, indicating that the hydrogen bonding between them is relatively weak. Furthermore, the peaks at 0.486 nm, 0.612 nm, 0.662 nm, and 0.455 nm indicate vdW interactions between the four drug molecules and GO. This result is consistent with the analysis in Figure 5b, which indicates that vdW interactions between the four drug molecules and GO still play a dominant role, while the hydrogen bonding is relatively weak.

The distribution of the angle between the aromatic ring of the drug molecule and the GO plane during kinetic adsorption was also analyzed; the results are shown in Figure 5d. The probability of the angular distribution between the aromatic rings of four drugs and the GO planes varied considerably compared to graphene during the simulation. Although the most probable angle between Gefi, CPT, Anas, and Res and GO remained relatively small, approximately 7°, 10°, 14°, and 9° respectively, the probability of occurrence decreased and, except for Res, there was a significant increase in the probability of the angle between the other three drug molecules and GO being greater than 30°. This also indicates that the π-π interactions between GO and drug molecules were weakened to some extent. This result mainly originates from the presence of oxygen-containing functional groups on GO.

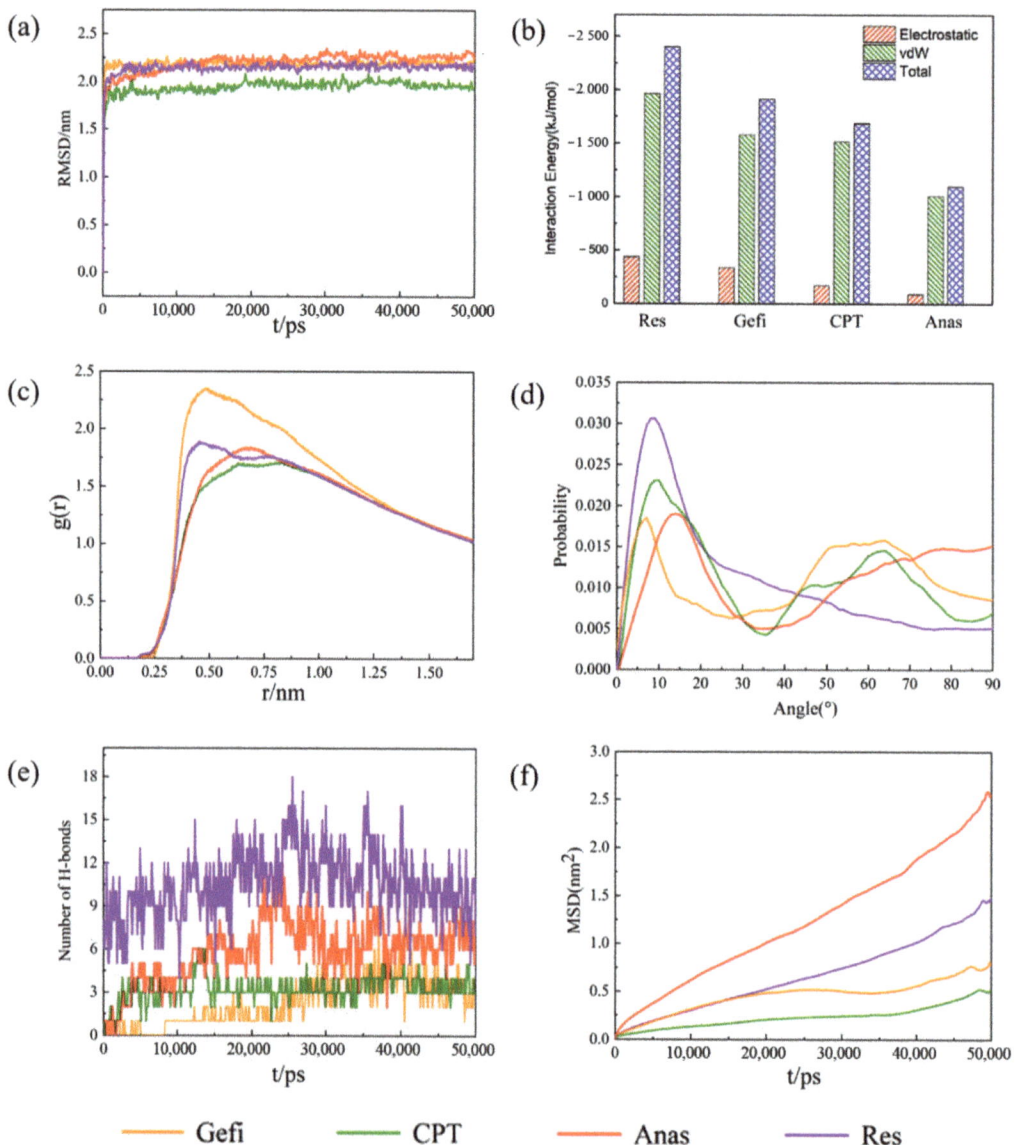

Figure 5. (**a**) The RMSD plots of the system with different drugs adsorbed on graphene as function of time (**b**) The average electrostatic, vdW, and total interaction energies of the systems with different drug molecules and GO. (**c**) Radial distribution functions (RDF) of the different drug molecules with GO. (**d**) The probability of the angle between the aromatic rings of the drug molecules and the GO plane. (**e**) Changes of the number of hydrogen bonds between the four kinds of drug molecules and GO over the time. (**f**) The MSD plots of the different drug molecules with GO.

The hydrogen bonding in the system also plays an important role in the overall adsorption process; therefore, the variation in the number of hydrogen bonds between different drug molecules and GO during the simulation was investigated, as shown in Figure 5e. Throughout the adsorption process, it can be seen that the number of hydrogen bonds formed between all four drug molecules and GO is relatively small, which is also consistent with the results obtained in Figure 5b. This further confirms that the hydrogen bonding between the four drug molecules and GO is relatively weak during the adsorption process.

To further understand the movement of molecules in the system, the diffusion coefficients of the four drug molecules were studied separately, Figure 5f shows the MSD plots of drug molecules adsorbed on GO. Table 2 shows the diffusion coefficients of drug molecules in different systems in descending order. Comparing with Table 1, it is found that the diffusion coefficients of all four drug molecules show different degrees of reduction, which also indicates that the adsorption of these four drugs on GO is slightly stronger than that on graphene [39,44,45].

Table 2. Self-diffusion coefficients of different drugs adsorbed on graphene oxide surface.

Drug Molecules	Gefi	CPT	Anas	Res
Diffusion coefficient ($\times 10^{-5}$ cm$^2\cdot$s^{-1})	0.001416	0.001032	0.007013	0.002307

4. Conclusions

In summary, we used DFT methods and MD simulations to investigate the adsorption processes and interaction mechanisms of graphene and GO with Gefi, CPT, Anas, and Res. From the result of DFT calculations, it is clear that GO has stronger adsorption properties than graphene for the four drug molecules, and the adsorption energy follows the order of Anas < CPT < Gefi < Res for both the graphene and GO systems. Regarding the adsorption systems of drug molecules with graphene oxide, static calculations further confirmed the preferential adsorption sites. By utilizing MD simulations, we found the adsorption mechanisms of different drugs with graphene as well as GO; it was found that π-π interactions and hydrogen bonding played an important role in the whole adsorption process. Among the four drug molecules, Res molecules showed the strongest adsorption capacity on graphene and GO, while Anas showed the weakest adsorption capacity on both graphene and GO. Furthermore, vdW interactions played a dominant role in the dynamic adsorption of drug molecules on both graphene and GO. Hydrogen-bonding had only a small contribution to the adsorption of drug molecules on GO. Taken together, GO has a stronger ability to adsorb these four drug molecules than graphene. Due to the good adsorption properties of graphene and GO for the four drug molecules, this study helps gain insight into the loading behavior of anti-cancer drugs on graphene, and also helps to provide assistance in the development of carriers of loaded drugs for anti-cancer drugs.

Author Contributions: Conceptualization, P.G. and Y.Z.; methodology, J.Z.; validation, Y.W. and Y.Z.; formal analysis, P.G.; investigation, H.L.; resources, Y.J.; data curation, P.G.; writing—original draft preparation, P.G.; writing—review and editing, P.G.; visualization, Y.Z.; supervision, P.Z.; project administration, Y.W.; funding acquisition, Y.J. All authors have read and agreed to the published version of the manuscript.

Funding: This research was funded by [the National Natural Science Foundation of China] grant number [Grant No. 52101287 and U1806219], [the Project of Taishan Scholar Construction Engineering and the program of the Jinan Science and Technology Bureau] grant number [2020GXRC019] and [the new material demonstration platform construction project from the Ministry of Industry and Information Technology] grant number [2020-370104-34-03-043952-01-11].

Institutional Review Board Statement: Not applicable.

Informed Consent Statement: Not applicable.

Data Availability Statement: Data is contained within article.

Acknowledgments: Special funding also supports this work from the Project of the Qilu Young Scholar Program of Shandong University.

Conflicts of Interest: The authors declare no conflict of interest.

Sample Availability: Not available.

References

1. Mostafavi, E.; Iravani, S. Mxene-graphene composites: A perspective on biomedical potentials. *Nano-Micro Lett.* **2022**, *14*, 130. [CrossRef] [PubMed]
2. Ji, Y.; Zhu, R.; Shen, Y.; Tan, Q.; Chen, J. Comparison of loading and unloading of different small drugs on graphene and its oxide. *J. Mol. Liq.* **2021**, *341*, 117454. [CrossRef]
3. Zhang, B.; Wei, P.; Zhou, Z.; Wei, T. Interactions of graphene with mammalian cells: Molecular mechanisms and biomedical insights. *Adv. Drug Deliv. Rev.* **2016**, *105*, 145–162. [CrossRef] [PubMed]
4. Oliveira, A.M.L.; Machado, M.; Silva, G.A.; Bitoque, D.B.; Tavares Ferreira, J.; Pinto, L.A.; Ferreira, Q. Graphene oxide thin films with drug delivery function. *Nanomaterials* **2022**, *12*, 1149. [CrossRef]
5. Fang, W.; Peng, L.; Liu, Y.; Wang, F.; Xu, Z.; Gao, C. A review on graphene oxide two-dimensional macromolecules: From single molecules to macro-assembly. *Chin. J. Polym. Sci.* **2021**, *39*, 267–308. [CrossRef]
6. Alinejad, A.; Raissi, H.; Hashemzadeh, H. Understanding co-loading of doxorubicin and camptothecin on graphene and folic acid-conjugated graphene for targeting drug delivery: Classical md simulation and dft calculation. *J. Biomol. Struct. Dyn.* **2020**, *38*, 2737–2745. [CrossRef]
7. Wall, M.E.; Wani, M.C.; Cook, C.; Palmer, K.; Mcphail, A.T.; Sim, G. Plant antitumor agents. I. The isolation and structure of camptothecin, a novel alkaloidal leukemia and tumor inhibitor from camptotheca acuminata[1,2]. *J. Am. Chem. Soc.* **1966**, *88*, 3888–3890. [CrossRef]
8. Thomas, C.J.; Rahier, N.J.; Hecht, S.M. Camptothecin: Current perspectives. *Bioorg. Med. Chem.* **2004**, *12*, 1585–1604. [CrossRef]
9. Jena, N.R.; Mishra, P.C. A theoretical study of some new analogues of the anti-cancer drug camptothecin. *J. Mol. Model.* **2006**, *13*, 267–274. [CrossRef]
10. Di Nunzio, M.R.; Cohen, B.; Douhal, A. Structural photodynamics of camptothecin, an anticancer drug in aqueous solutions. *J. Phys. Chem. A* **2011**, *115*, 5094–5104. [CrossRef]
11. Liu, S.; Yuan, Y.; Okumura, Y.; Shinkai, N.; Yamauchi, H. Camptothecin disrupts androgen receptor signaling and suppresses prostate cancer cell growth. *Biochem. Biophys. Res. Commun.* **2010**, *394*, 297–302. [CrossRef] [PubMed]
12. Gomes, B.A.Q.; Queiroz, A.N.; Borges, R.S. A theoretical study of resveratrol oxidation. *J. Comput. Theor. Nanosci.* **2009**, *6*, 1637–1639. [CrossRef]
13. Stivala, L.A.; Savio, M.; Carafoli, F.; Perucca, P.; Bianchi, L.; Maga, G.; Forti, L.; Pagnoni, U.M.; Albini, A.; Prosperi, E.; et al. Specific structural determinants are responsible for the antioxidant activity and the cell cycle effects of resveratrol. *J. Biol. Chem.* **2001**, *276*, 22586–22594. [CrossRef] [PubMed]
14. Zhou, J.H.; Cheng, H.Y.; Yu, Z.Q.; He, D.H.; Pan, Z.; Yang, D.T. Resveratrol induces apoptosis in pancreatic cancer cells. *Chin. Med. J.* **2011**, *124*, 1695–1699.
15. Yokouchi, H.; Yamazaki, K.; Kinoshita, I.; Konishi, J.; Asahina, H.; Sukoh, N.; Harada, M.; Akie, K.; Ogura, S.; Ishida, T.; et al. Clinical benefit of readministration of gefitinib for initial gefitinib-responders with non-small cell lung cancer. *BMC Cancer* **2007**, *7*, 51. [CrossRef]
16. Li, J.; Kleeff, J.; Giese, N.; Buchler, M.W.; Korc, M.; Friess, H. Gefitinib ('iressa', zd1839), a selective epidermal growth factor receptor tyrosine kinase inhibitor, inhibits pancreatic cancer cell growth, invasion, and colony formation. *Int. J. Oncol.* **2004**, *25*, 203–210. [CrossRef]
17. Geng, D.; Sun, D.; Zhang, L.; Zhang, W. Materials and methods the statement of ethics. *Afr. Health Sci.* **2015**, *15*, 594–597. [CrossRef]
18. Kalykaki, A.; Agelaki, S.; Kallergi, G.; Xyrafas, A.; Mavroudis, D.; Georgoulias, V. Elimination of egfr-expressing circulating tumor cells in patients with metastatic breast cancer treated with gefitinib. *Cancer Chemother. Pharm.* **2014**, *73*, 685–693. [CrossRef]
19. Reese, D.; Nabholtz, J.M. Anastrozole in the management of breast cancer. *Expert Opin. Pharmacother.* **2002**, *3*, 1329–1339. [CrossRef]
20. Cuzick, J.; Sestak, I.; Forbes, J.F.; Dowsett, M.; Cawthorn, S.; Mansel, R.E.; Loibl, S.; Bonanni, B.; Evans, D.G.; Howell, A. Use of anastrozole for breast cancer prevention (ibis-ii): Long-term results of a randomised controlled trial. *Lancet* **2020**, *395*, 117–122. [CrossRef]
21. Nabholtz, J. Role of anastrozole across the breast cancer continuum: From advanced to early disease and prevention. *Oncology* **2006**, *70*, 1–12. [CrossRef] [PubMed]
22. Graham, P.H. Anastrozole for malignant and benign conditions: Present applications and future therapeutic integrations. *Expert Opin. Pharmacother.* **2007**, *8*, 2347–2357. [CrossRef] [PubMed]

23. Ayissi, S.; Charpentier, P.A.; Farhangi, N.; Wood, J.A.; Palotas, K.; Hofer, W.A. Interaction of Titanium Oxide Nanostructures with Graphene and Functionalized Graphene Nanoribbons: A DFT Study. *J. Phys. Chem. C* **2013**, *117*, 25424–25432. [CrossRef]
24. Perdew, J.P.; Burke, K.; Ernzerhof, M. Generalized gradient approximation made simple. *Phys. Rev. Lett.* **1996**, *77*, 3865–3868. [CrossRef] [PubMed]
25. Delley, B. An all-electron numerical method for solving the local density functional for polyatomic molecules. *J. Chem. Phys.* **1990**, *92*, 508–517. [CrossRef]
26. Delley, B. From molecules to solids with the dmol3 approach. *J. Chem. Phys.* **2000**, *113*, 7756–7764. [CrossRef]
27. Steinmann, S.N.; Csonka, G.; Corminboeuf, C. Unified inter- and intramolecular dispersion correction formula for generalized gradient approximation density functional theory. *J. Chem. Theory Comput.* **2009**, *11*, 2950. [CrossRef]
28. Janesko, B.G.; Barone, V.; Brothers, E.N. Accurate surface chemistry beyond the generalized gradient approximation: Illustrations for graphene adatoms. *J. Chem. Theory Comput.* **2013**, *11*, 4853–4859. [CrossRef]
29. Hill, T.L. Statistical mechanics of multimolecular adsorption. I. *J. Chem. Phys.* **1946**, *14*, 263–267. [CrossRef]
30. Hunter, C.A.; Sanders, J.K. The nature of tt-tt interactions. *J. Am. Chem. Soc.* **1990**, *112*, 5525–5534. [CrossRef]
31. Kazachenko, A.S.; Akman, F.; Sagaama, A.; Issaoui, N.; Malyar, Y.N.; Vasilieva, N.Y.; Borovkova, V.S. Theoretical and experimental study of guar gum sulfation. *J. Mol. Model.* **2021**, *27*, 5. [CrossRef] [PubMed]
32. Pitoňák, M.; Neogrády, P.; Řezáč, J.; Jurečka, P.; Urban, M.; Hobza, P. Benzene dimer: High-level wave function and density functional theory calculations. *J. Chem. Theory Comput.* **2008**, *4*, 1829–1834. [CrossRef] [PubMed]
33. Gao, H.; Liu, Z. Dft study of no adsorption on pristine graphene. *RSC Adv.* **2017**, *7*, 13082–13091. [CrossRef]
34. Wang, H.; Jia, Y. Udmh adsorption on graphene oxides: A first-principles study. *Diam. Relat. Mater.* **2021**, *117*, 108457. [CrossRef]
35. Kozlov, S.M.; Viñes, F.; Görling, A. On the interaction of polycyclic aromatic compounds with graphene. *Carbon* **2012**, *50*, 2482–2492. [CrossRef]
36. Zhao, D.; Li, L.; Zhou, J. Simulation insight into the cytochrome c adsorption on graphene and graphene oxide surfaces. *Appl. Surf. Sci.* **2018**, *428*, 825–834. [CrossRef]
37. Maleki, R.; Khoshoei, A.; Ghasemy, E.; Rashidi, A. Molecular insight into the smart functionalized tmc-fullerene nanocarrier in the ph-responsive adsorption and release of anti-cancer drugs. *J. Mol. Graph. Model.* **2020**, *100*, 107660. [CrossRef]
38. Kuisma, E.; Hansson, C.F.; Lindberg, T.B.; Gillberg, C.A.; Idh, S.; Schröder, E. Graphene oxide and adsorption of chloroform: A density functional study. *J. Chem. Phys.* **2016**, *144*, 184704. [CrossRef]
39. Ai, Y.; Liu, Y.; Huo, Y.; Zhao, C.; Sun, L.; Han, B.; Cao, X.; Wang, X. Insights into the adsorption mechanism and dynamic behavior of tetracycline antibiotics on reduced graphene oxide (rgo) and graphene oxide (go) materials. *Environ. Sci. Nano* **2019**, *6*, 3336–3348. [CrossRef]
40. Zhao, M.; Lai, Q.; Guo, J.; Guo, Y. Insights into the adsorption of resveratrol on graphene oxide: A first-principles study. *ChemistrySelect* **2017**, *2*, 6895–6900. [CrossRef]
41. Kamel, M.; Raissi, H.; Hashemzadeh, H.; Mohammadifard, K. Theoretical elucidation of the amino acid interaction with graphene and functionalized graphene nanosheets: Insights from dft calculation and md simulation. *Amino Acids* **2020**, *52*, 1465–1478. [CrossRef] [PubMed]
42. Tang, H.; Zhao, Y.; Shan, S.; Yang, X.; Liu, D.; Cui, F.; Xing, B. Theoretical insight into the adsorption of aromatic compounds on graphene oxide. *Environ. Sci. Nano* **2018**, *5*, 2357–2367. [CrossRef]
43. Headen, T.F.; Howard, C.A.; Skipper, N.T.; Wilkinson, M.A.; Bowron, D.T.; Soper, A.K. Structure of π–π interactions in aromatic liquids. *J. Am. Chem. Soc.* **2010**, *132*, 5735–5742. [CrossRef] [PubMed]
44. Yao, N.; Li, C.; Yu, J.; Xu, Q.; Wei, S.; Tian, Z.; Yang, Z.; Yang, W.; Shen, J. Insight into adsorption of combined antibiotic-heavy metal contaminants on graphene oxide in water. *Sep. Purif. Technol.* **2020**, *236*, 116278. [CrossRef]
45. Tang, H.; Zhang, S.; Huang, T.; Cui, F.; Xing, B. Ph-dependent adsorption of aromatic compounds on graphene oxide: An experimental, molecular dynamics simulation and density functional theory investigation. *J. Hazard. Mater.* **2020**, *395*, 122680. [CrossRef] [PubMed]

Article

Graphene@Curcumin-Copper Paintable Coatings for the Prevention of Nosocomial Microbial Infection

Mohammad Oves [1], Mohammad Omaish Ansari [2], Mohammad Shahnawaze Ansari [2] and Adnan Memić [2,*]

[1] Center of Excellence in Environmental Studies, King Abdulaziz University, Jeddah 21589, Saudi Arabia
[2] Center of Nanotechnology, King Abdulaziz University, Jeddah 21589, Saudi Arabia
* Correspondence: amemic@kau.edu.sa; Tel.: +966-564975404

Abstract: The rise of antimicrobial resistance has brought into focus the urgent need for the next generation of antimicrobial coating. Specifically, the coating of suitable antimicrobial nanomaterials on contact surfaces seems to be an effective method for the disinfection/contact killing of microorganisms. In this study, the antimicrobial coatings of graphene@curcumin-copper (GN@CR-Cu) were prepared using a chemical synthesis methodology. Thus, the prepared GN@CR-Cu slurry was successfully coated on different contact surfaces, and subsequently, the GO in the composite was reduced to graphene (GN) by low-temperature heating/sunlight exposure. Scanning electron microscopy was used to characterize the coated GN@CR-Cu for the coating properties, X-ray photon scattering were used for structural characterization and material confirmation. From the morphological analysis, it was seen that CR and Cu were uniformly distributed throughout the GN network. The nanocomposite coating showed antimicrobial properties by contact-killing mechanisms, which was confirmed by zone inhibition and scanning electron microscopy. The materials showed maximum antibacterial activity against *E. coli* (24 ± 0.50 mm) followed by *P. aeruginosa* (18 ± 0.25 mm) at 25 µg/mL spot inoculation on the solid media plate, and a similar trend was observed in the minimum inhibition concentration (80 µg/mL) and bactericidal concentration (160 µg/mL) in liquid media. The synthesized materials showed excellent activity against *E. coli* and *P. aeruginosa*. These materials, when coated on different contact surfaces such medical devices, might significantly reduce the risk of nosocomial infection.

Keywords: antimicrobial; coating; graphene; curcumin; copper; *Pseudomonas aeruginosa*; *E. coli*

Citation: Oves, M.; Ansari, M.O.; Ansari, M.S.; Memić, A. Graphene@Curcumin-Copper Paintable Coatings for the Prevention of Nosocomial Microbial Infection. *Molecules* 2023, 28, 2814. https://doi.org/10.3390/molecules28062814

Academic Editors: Minas M. Stylianakis and Athanasios Skouras

Received: 7 February 2023
Revised: 2 March 2023
Accepted: 7 March 2023
Published: 20 March 2023

Copyright: © 2023 by the authors. Licensee MDPI, Basel, Switzerland. This article is an open access article distributed under the terms and conditions of the Creative Commons Attribution (CC BY) license (https://creativecommons.org/licenses/by/4.0/).

1. Introduction

Infectious diseases that are caused by microorganisms can lead to serious health complications, including death, for both humans and animals. The attachment of microbes to surfaces leading to the formation of biofilms poses a particularly serious threat. Many industries, ranging from healthcare systems, the food and water industry, as well as the oil and gas industry, suffer huge losses due to complications as a result of biofilm formation [1]. Further challenges arise when considering that microbes adhering to the surfaces can be transferred when touched, thereby spreading microbial colonization. During the global COVID-19 outbreak, the development of antimicrobial surfaces attracted the attention of scientists worldwide. The development of antimicrobial coating could be an effective strategy to prevent microbial spread in general and specifically biofilm formation. [2]. To achieve these goals, several approaches have been proposed [3]. For example, recently, a combination of different nanomaterials, such as graphene (GN), metals and metal oxides, or natural antibacterial materials, have been proposed as antimicrobial coatings. Such a combination of materials could have exciting synergistic or additional effects when compared to the individual components developed for anti-infective surface deployment.

Some recently designed GN-based nanomaterials are known to exhibit antimicrobial activity towards bacteria [4], fungi [5], and viruses [6]. The direct physicochemical interaction between the GN-based materials and microbes can lead to the damage of the cellular components, principally the proteins, lipids, and nucleic acids [7]. The high affinity of GN-based materials towards membrane proteoglycans leads to membrane damage. Another mechanism of action is the further internal leakage of GN-based materials into cells which can lead to the inhibition of DNA/RNA replication [8]. Therefore, it is proposed that the mode of action of GN towards bacteria is by physical means as well as by chemical means [9]. The physical means operate by the direct contact of microorganisms with the sharp edges of GN. For example, Akhavan et al. [10] showed that the GN-based materials could potentially encapsulate microorganisms for their isolation but also act as an effective photothermal agent for the inactivation of the GN-wrapped microorganisms. Pham et al. [11] similarly showed that the density of GN edges significantly affects its antibacterial activity. The researchers proposed that GN sheets act as blades cutting through the cell membrane and induce pores, leading to their rupture leading to bacterial death. On the other hand, chemical mechanisms involve oxidative stress and the generation of reactive oxygen species (ROS) responsible for the inactivation of microorganisms [12].

Among metallic particles, many research groups have shown that copper (Cu) and its oxides have antimicrobial properties [13–20]. The mechanism of killing involves bacterial cell wall damage leading to the loss of cytoplasmic content [21]. Apart from this, the reactive oxygen species can induce even greater damage to organelles and even lead to nucleic acid degradation [22]. Cu has also been found to possess high antiviral properties as well [23]. Bleichert et al. [24] showed that attenuated vaccine strain vaccinia virus and virulent MPXV Copenhagen were inhibited within 3 min of exposure to Cu. Similarly, Noyce et al. [25] showed that the Cu alloys (61–95% Cu) effectively killed *E. coli* at room temperature. They also found that samples with a high percentage of Cu possessed the highest antibacterial properties. Similar results were also reported by Wilks et al. [26], who showed that a Cu content of 85% or more showed good antibacterial activities. Similar effects of Cu were found against the vegetative and spore forms of *Clostridium difficile* and a significant reduction in survival of the *C. difficile* vegetative cells and spores was observed after 24–48 h [27]. From the above discussion, it appears that the amount of Cu can have a direct effect on the contact killing of microorganisms, including both bacteria and viruses.

Among naturally occurring antimicrobial materials, curcumin (CR) has been widely used since its antibacterial properties were demonstrated by Schraufstatter and Bernt in 1949 [28]. CR promotes recombinant protein overexpression, thereby leading to an apoptosis-like response in bacteria [29]. Several investigations revealed that CR had antibacterial effects on both Gram-positive and Gram-negative bacteria [30]. CR antibacterial activity includes bacterial membrane rupture, the suppression of virulence factor synthesis and suppression of biofilm formation, and the activation of oxidative stress [31,32]. Recently, Oves et al. [33] has investigated CR- and ZnO-glazed GN for the successful growth inhibition of a Methicillin-resistant bacterial strain of *Staphylococcus aureus*.

Due to the antibacterial properties of both GN and Cu paired with CR, it can be interpreted that their combination will be highly effective in combating different types of bacteria via contact killing. Thus, in this work, solutions of Cu particles and CR dispersed in graphene oxide (GO) gel was prepared. The prepared dispersion was applied on contact surfaces and its subsequent heat treatment resulted in the reduction of GO into reduced-graphene oxide (rGO) (Figure 1). The Cu, CR, and reduced-graphene oxide coatings were thereafter tested for the contact killing of *Pseudomonas aeruginosa* (*P. aeruginosa*) and *Escherichia coli* (*E. coli*).

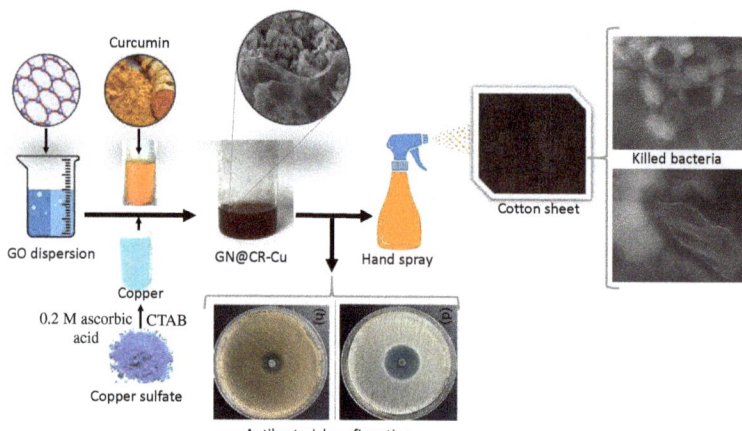

Figure 1. Schematic representation of the synthesis and antibacterial application of GN@CR-Cu.

2. Results and Discussion

2.1. Morphological Analysis

FESEM images of GN, GN@Cu, GN@CR, and GN@CR-Cu at low and high magnifications (insets) are shown in Figure 2a–d. The folded sheet-type structure can be very clearly observed in Figure 2a, which is a known feature of GN. When it comes to GN@Cu (Figure 2b), solid polyhedral-shaped Cu particles are visible below and above the GN sheets. It is evident from the inset of Figure 2a that the Cu particles are a bit blurry because of their presence underneath a very thin and nearly transparent layer of GN. On the other hand, the well-dispersed fluffy circular geometries of the CR particles with GN sheets can be observed in Figure 2c. The CR particles are easily detectable in a small void between the single and multi-layered GN sheets in the inset of Figure 2c. As far as the GN@CR-Cu composite is concerned (Figure 2d), the sample is homogeneous with Cu, and CR particles are uniformly distributed. The inset of Figure 2d also reveals that the polyhedral solid geometries of the Cu particles and slightly circular-shaped CR particles are sandwiched between the ultrafine layers of GN. The FESEM image of CR and Cu nanoparticles can be seen in Figure S1.

Figure 3a–e shows the elemental mapping and quantitative analysis of the as-synthesized GN@CR-Cu sample recorded by using EDS. The mixed electron image of the GN@CR-Cu (Figure 3a), C in red (Figure 3b), O in green (Figure 3c), and Cu in blue shows that the prepared sample, in the form of a composite, has all three (GN, Cu, and CR) components with uniform distribution. Figure 3e illustrates the EDS spectrum consisting of C, O, and Cu peaks, indicating that the GN@CR-Cu composite was successfully fabricated.

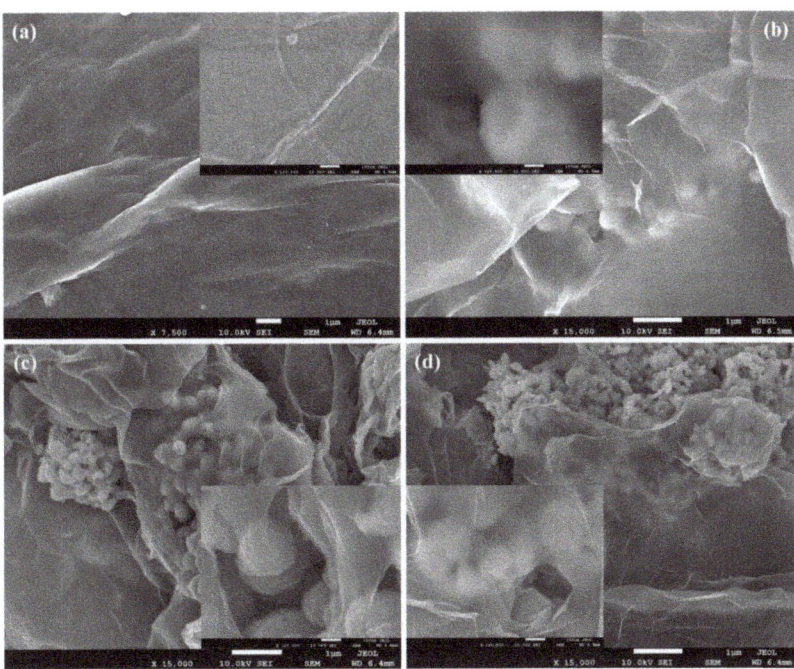

Figure 2. Low and high magnification (insets) FESEM images of (**a**) GN; (**b**) GN@Cu; (**c**) GN@CR; and (**d**) GN@CR-Cu.

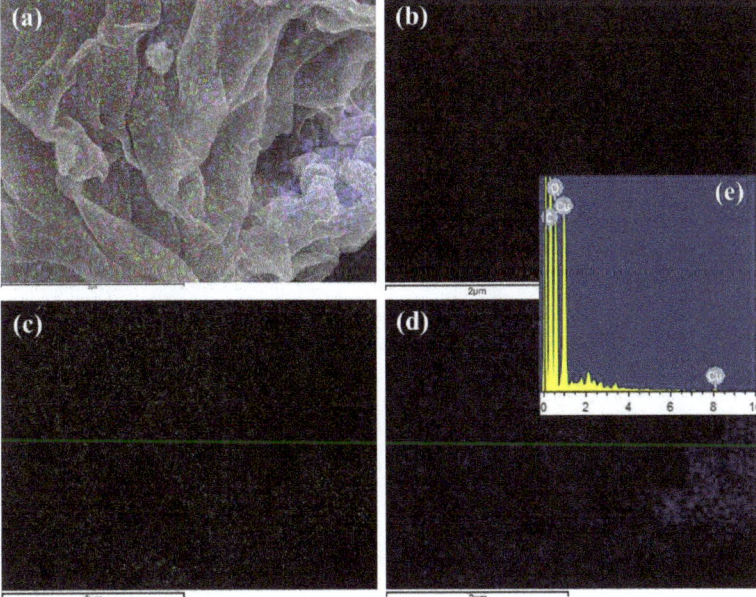

Figure 3. Elemental mapping images of (**a**) GN@CR-Cu; (**b**) C; (**c**) O; (**d**) Cu; and (**e**) EDS spectrum of GN@CR-Cu.

Figure 4a,b shows low and high magnification HRTEM images of GN@CR-Cu. Figure 4a reveals that the sample is well-formed and homogeneous with a uniform amalgamation of GN, Cu, and CR. All three components, the sheet-like layered structure of GN, the almost hexagonal/polyhedral-shaped solid surface of Cu, and the fluffy/circular geometry of CR particles can be easily identified in the image Figure 4b.

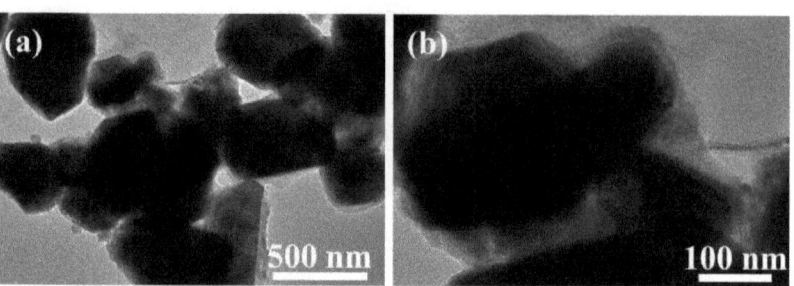

Figure 4. (a) Low magnification; (b) High magnification HRTEM images of GN@CR-Cu.

2.2. X-ray Photoelectron Spectroscopy (XPS)

The XPS analysis was done to study the elements present, including possible impurities in the Cu, GN, and GN@CR-Cu coatings. In the case of pure Cu, peaks corresponding to C1s, O1s, and Cu2p3 are observed (Figure S2). The presence of a small percentage of O1s is due to the slight oxidation of Cu upon photothermal treatment. In the case of GN, peaks corresponding to C1s and O1s are present. The O1s depict small functionalization or a slightly unreduced part of GO. The survey scan of GN@CR-Cu showed the peaks corresponding to C1s, O1s, and Cu2p3, but here, in contrast to the spectra of pure Cu, a high percentage of C1s and Cu2p3 was observed (Figure 5). The C1s peak can be deconvoluted into three peaks at 284.8, 286.1, and 288.3 eV, corresponding to the sp2 and sp3 (C=C/C–C) bonding, C-O chemical bonds, and carbonyl groups (C=O/COO) in rGO [34–36]. The Cu2p peak that can be deconvoluted into the two peaks at 935.1 and 934.8 eV were attributed to the Cu^{2+} and Cu^+ chemical states [37]. The peak at 942.5 eV is due to the shakeup satellite peak [38].

2.3. Antibacterial Performance of GN@CR-Cu Composite

In this study, the antimicrobial testing of GN, GN@CR, GN@Cu, and GN@CR-Cu was conducted against *E. coli* and *P. aeruginosa* bacterial strains. Among these, the GN@CR-Cu composite material showed highly efficient antimicrobial activities against both nosocomial bacterial strains. The *E. coli* bacterial strain was more significantly affected as compared to the *P. aeruginosa* bacterial strain in both assays. The nanocomposite can act by bypassing drug resistance mechanisms in bacteria and inhibiting biofilm formation or other important processes related to their virulence potential [39]. Nanoparticles can penetrate the cell wall and membrane of the bacteria and disrupt important molecular mechanisms [40]. In general, *E. coli* is a facultative anaerobic bacterial species, while *P. aeruginosa* is an aerobic bacterial species. *P. aeruginosa* promoting *E. coli* biofilm formation in a nutrient-limited medium has been reported earlier [41]. *P. aeruginosa* can produce more exopolysaccharides than *E. coli*. These *P. aeruginosa* exopolysaccharides play an important role in biofilm formation, and due to it, *P. aeruginosa* rapidly forms biofilm compared to *E. coli*. Therefore, this exopolysaccharide inhibits the binding of nanomaterial of the *P. aeruginosa* bacterial cell membrane and is hypothesized to contribute to being less affected when compared to *E. coli*. The bacterial strain of *E. coli* and *P. aeruginosa* growth was significantly influenced and developed a bacterial growth inhibition zone around the nanocomposite material, where the compound diffused into the surrounding media. The GN@CR-Cu showed a significant zone of inhibition of 24 ± 0.50 and 18 ± 0.25 mm against *E. coli* and *P. aeruginosa* at 25 µg/mL, while GN@Cu showed 18 ± 0.25 and 14 ± 0.5 mm against

E. coli and *P. aeruginosa* at a 25 μg/mL concentration, and GN@CR showed 17 ± 0.75 and 12 ± 0.5 mm zone inhibition against *E. coli* and *P. aeruginosa* at the 25 μg/mL concentrations, respectively (Figure 6).

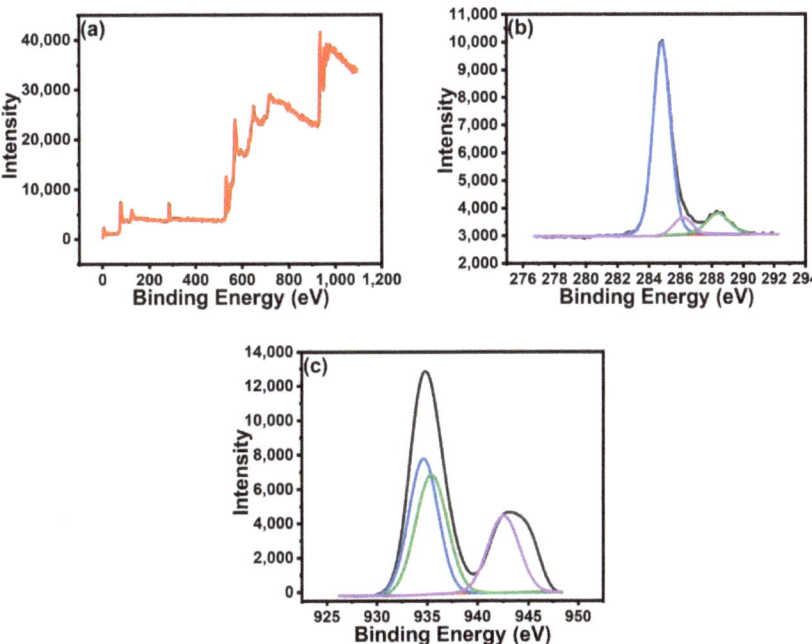

Figure 5. XPS spectra of GN@CR-Cu composite: (**a**) survey scan, (**b**) C1s, and (**c**) Cu2p3.

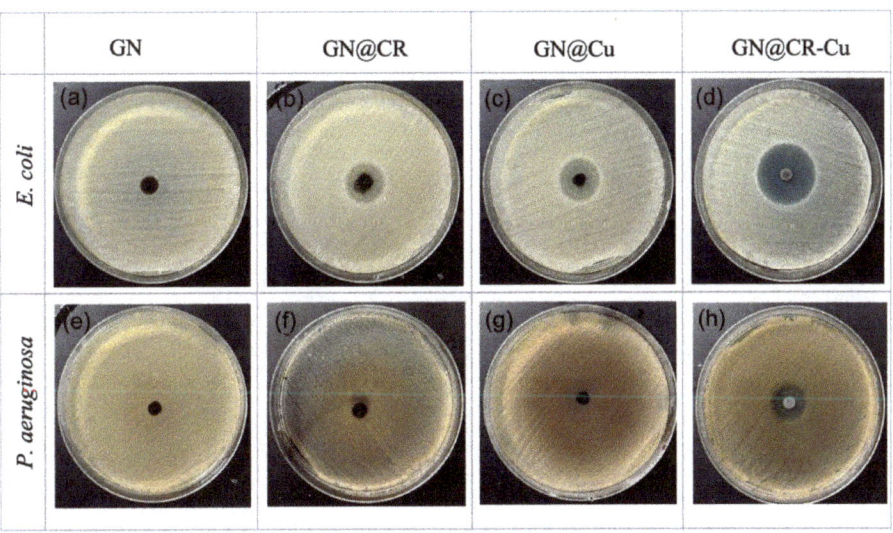

Figure 6. Test bacteria *E. coli* (**a**–**d**), *P. aeruginosa* (**e**–**h**): The zone inhibition by the nanocomposite material on the bacteria cultivated on nutrient agar media plates.

In addition, antibacterial activity was examined by testing it in a broth and determining its minimum inhibitory concentrations (MIC) and minimum bactericidal concentrations (MBC). It was reported that the inhibitory concentration against both microorganisms individually may reach up to 80, 160, and 160 μg/mL for GN@CR-Cu, GN@CR, and GN@Cu, respectively. The antibacterial activity of the composites was evaluated by putting it into a liquid culture nutrient broth along with the inoculated test bacteria and subsequently inoculating it as described in Section 3.2.1. After overnight incubation, a clear pattern of bacterial growth was found on the nutrient agar plate, which is shown in Figure 7. Further antimicrobial work was performed in terms of bacterial survivability in the presence of composite materials (Figure 8). GN alone did not significantly inhibit bacterial growth in the cases of both *E. coli* and *P. aeruginosa*. The GN composites with either CR or Cu were effective antibacterial agents; however, their activity was less potent when compared to the GN@CR-Cu composite (Figure 8a,b).

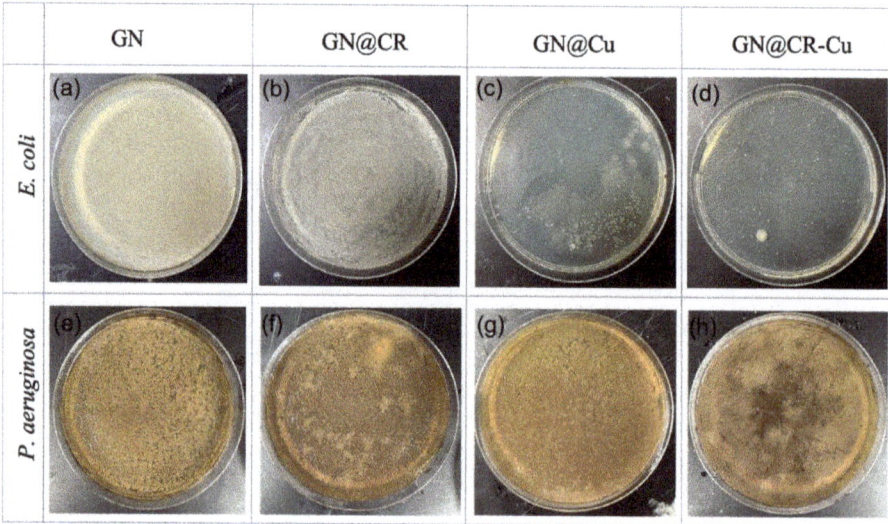

Figure 7. Test bacteria *E. coli* (**a–d**), *P. aeruginosa* (**e–h**): The minimum inhibition by the nanocomposite material on the bacteria cultivated Petri plates.

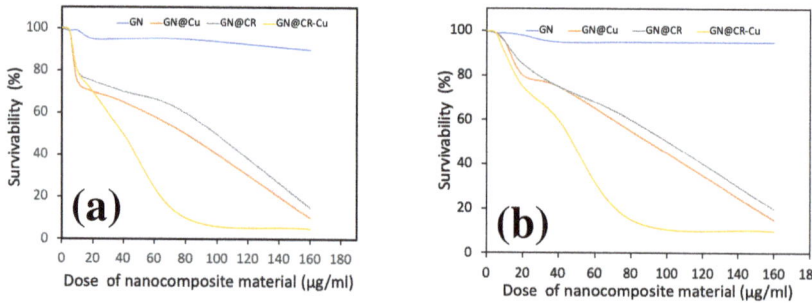

Figure 8. The survivability of *E. coli* (**a**) and *P. aeruginosa* (**b**) in the presence of nanocomposite material. Image showing excellent dose-response, increasing concentration significantly retards bacterial growth.

We hypothesize that the GN@CR-Cu antibacterial mechanism combines both chemical and physical modes of action. GN enriched with the chemical agent Cu and biological or natural antimicrobial agent CR showed superior antibacterial activity when compared to other formulations (i.e., GN, GN@CR, or GN@Cu). We hypothesize that when the nanocomposite material encounters bacterial cells, the sharp edge of the GN pierce the bacterial membranes causing the leakage of cell organelles. Similarly, when the nanocomposite is coated on solid surfaces, the water contact angle measured ~70° (Figure S3). We hypothesize that the hydrophilic surface might contribute to the decreased bacterial attachment ability of the strains tested [42,43]. It might also be possible that the creation of various active oxygen species occurs, which prevents the growth of bacterial cells due to the presence of Cu in the formulation [44]. In the previously published studies, it was demonstrated that Cu NPs have outstanding antibacterial activity. In a recent study, the Cu nanoparticle size and concentration have a direct effect on the antimicrobial activity of *E. coli* via numerous mechanisms [45].

In this study, GN deposited with Cu nanomaterial significantly enhances its antimicrobial activity clearly shown in the zone inhibition (Figures 6–8) and electron microscopy image (Figure 9) studies. This material was further enriched with the addition of natural CR, which reveals excellent antimicrobial activity. Recently, the antibacterial potential of bulk CR and nano CR against the *Staphylococcus aureus* and *E. coli* was successfully investigated by Sandhuli et al. [46]. The inhibition zones of the nano-formulated CR cream were greater than those of bulk CR cream for both *S. aureus* and *E. coli*, demonstrating its superior antibacterial action [46]. In our case, the GN@CR-Cu showed better antimicrobial potential than GN@Cu and GN@CR alone, most likely due to the synergistic effect of the two antimicrobial agents [47].

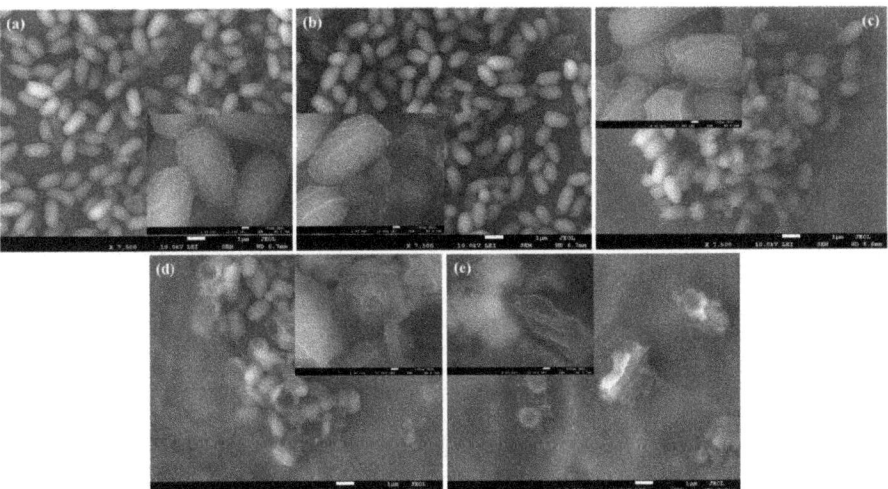

Figure 9. Scanning electron microscopy image of *E. coli* without any treatment as a control without an effect on cell morphology (**a**), and *E. coli* culture treated with GN (**b**), GN@CR (**c**), partial cell damage by the treatment of GN@Cu (**d**), and complete bacterial cell damage by the treatment of GN@CR-Cu (**e**).

The antibacterial mechanisms involved in the bactericidal activity of GN-containing nanomaterials, in general, could be summarized using the following mechanisms: (i) physical direct interaction of the extremely sharp edges of nanomaterials against the cell wall membrane [48], which can show further stress under sonication, (ii) ROS generation [49] even in the dark [50], (iii) trapping cells within the aggregated nanomaterials [7], (iv) oxidative stress [51], (v) interruption in the glycolysis process of the bacteria [52], (vi) DNA damage [53], (vii) metal ion release [54], and recently, (viii) contribution in generation/explosion of nanobubbles [55]. In the case of our nanocomposite, we hypothesize that the multiple modes of actions are at play, including bacterial cell damage that occurs due to the nanoparticle interaction. GN@Cu-CR materials are multifunctional due to the metal presence and natural CR. In order to prevent microbes from attaching, colonizing, spreading, and creating biofilms in medical devices, composite materials have the potential for external uses as antibacterial agents in the surface coatings on a variety of substrates.

Here, the nanocomposite material has been synthesized with a highly stable material, i.e., GN which stabilizes both CR and Cu. Due to the high stability of GN-based materials, the CR and Cu molecules fixed in the GN groves retain their antimicrobial activity. In our previous reports, the GN-based nanocomposite material with Zinc oxide and CR showed similarly excellent stability and antimicrobial activity against the multidrug-resistant *Staphylococcus aurous* bacterial strain [33].

3. Materials and Methods

3.1. Materials

Copper sulphate pentahydrate $CuSO_4 \cdot 5H_2O$ was sourced from Fluka (Buchs, Switzerland), and cetyltrimethylammonium bromide (CTAB) from Otto chemicals (Mumbai, India). Sulphuric acid, phosphoric acid, potassium permanganate, and ascorbic acid ($C_6H_8O_6$) were obtained from Sigma Aldrich (St. Louis, MO, USA). CR was purchased from a local supermarket of Jeddah, Saudi Arabia, and was dried under the Sun and subsequently crushed into fine powder. The water used in the experiments was deionized water.

3.1.1. Synthesis of Cu and GO@CR-Cu Suspension

The Cu nanoparticles were synthesized by the reduction of copper (II) sulfate in the presence of CTAB surfactant. In a typical process, 0.1 M copper (II) sulfate solution was dissolved in 100 mL of water, and to it, 0.25 g of CTAB was added and the whole system was put under stirring conditions. In another beaker, 50 mL of 0.2 M ascorbic acid solution was prepared. In the second step, the solution of ascorbic acid was slowly added to the copper (II) sulfate solution, and subsequently, 30 mL of 1 M sodium hydroxide solution was also added. The whole system was heated to 80 °C for 2 h and a dark reddish-brown color confirmed the formation of Cu. Thus, the prepared Cu was separated by centrifugation, washed with an excess of water and ethanol, and subsequently dried at room temperature [56].

GO was prepared using the modified Tour's method. In a typical process, to a 9:1 stirring mixture of concentrated H_2SO_4/H_3PO_4 (360:40 mL), 18 gms of $KMnO_4$ followed by 3 g of graphite flakes was added slowly. The whole system was left stirring at room temperature for 72 h for the exfoliation of graphite flakes. Thereafter, the mixture was cooled by putting it in an ice bath and was then subsequently poured into another beaker containing 400 mL of deionized water ice. To this cooled solution, 1–3 mL 30% H_2O_2 was added until the appearance of a yellow color confirmed the formation of GO. Thus, the synthesized GO was separated by centrifugation/filtration, washed to 100 mL of 5% HCl, excess of water, ethanol, and subsequently stored as a gel [57]. The GO solution was optimized and a stock solution of 10 mg/mL was prepared for future use.

3.1.2. Preparation of GO@CR-Cu Slurry

To 50 mL of the GO (10 mg/mL) solution, 0.125 g of both Cu and CR was added. The whole system was put on ultrasonic bath and later kept on stirring for the uniform distribution of Cu and CR in the GO. Just before coating, the mixture was repeatedly stirred/shaken for the uniform distribution of Cu and CR. For the reduction of GO inside GO@CR-Cu, the coated GO@CR-Cu was kept in sunlight on a bright sunny day for 8 h from 8 a.m. to 4 p.m. in Jeddah. For the evaluation of antimicrobial activity, the above GO@CR-Cu composite was taken in a Petri plate, reduced into GN@CR-Cu by the process described earlier, and finally used in its powdered form.

3.1.3. Characterization

The morphological and compositional analysis of GN@CR-Cu were conducted by field emission scanning electron microscopy (FESEM) (JSM-7600F from JEOL, Tokyo, Japan). The elemental mapping was recorded using the energy dispersive X-ray spectroscope (EDS) from Oxford Instruments, Oxfordshire, UK equipped with FESEM. For the elemental detection, X-ray photoelectron spectroscopy (XPS) (ESCALAB 250 from Thermo Fisher Scientific, Warrington, UK) was used at a monochromatized Al Kα X-ray source λ 1/4 1486.6 eV. The antibacterial studies were performed by studying the surface growth inhibition assay, and the minimum inhibitory concentration (MIC) and minimum bactericidal concentration (MBC) were determined against the *Escherichia coli* and *Pseudomonas aeruginosa* microorganisms.

3.2. Assessment of Antibacterial Activity of GN@CR-Cu Composite

3.2.1. Zone Inhibition Assay

The two bacterial strains, *Escherichia coli*, and *Pseudomonas aeruginosa* are frequently linked to nosocomial infections. This experiment investigated the antibacterial capability of materials that were synthesized. To obtain the best growth acquisition, these bacterial cultures were first grown in a Luria Bertani broth, and then a loopful culture was injected into the 100 mL liquid broth and incubated in a rotatory incubator. A fresh culture of each strain was created by re-culturing with the same medium under the ideal circumstances. The nutrition agar plate was created by utilizing the pour plate technique after the medium was autoclaved for 20 min in a sterilizer machine at the proper temperature of 121 °C and pressure of 15 pounds per square inch (JSR Autoclave, JSAC-80, Gongju-Cit, Republic of Korea). The media was poured into the Petri plates and solidified after 10 min of sitting. Each bacterial strain was moved from the fresh culture tubes to the media plates and then left there for 10 min to maintain the conditions the medium on the plate afforded. A total of 25 micrograms of the synthesized composite materials were added to the surface of each media plate, which previously contained bacteria. The nanocomposite material came into contact with the bacteria and spread into the surrounding media after an overnight incubation at the ideal temperature of 37 °C, revealing the antibacterial activity around the composite materials and developing a clear zone of bacterial growth inhibition.

3.2.2. MIC and MBC of GN@CR-Cu Composite

The effect of GN@CR-Cu nanocomposite on the growth of bacteria in broth was investigated, and the lowest concentrations required to impede growth (MIC) and kill bacteria (MBC) were determined. The GN@CR-Cu nanocomposite was used as an antibacterial agent against specific test microorganisms after being suspended in Milli-Q water. The broth dilution process is the approach that is the industry standard for assessing bacterial survival in the presence of the tested test agent or nanocomposite material. The concentration was adjusted by adding nanomaterials from 0 to 100 µg/mL. In the beginning, the same techniques were followed to set up twelve 500 mL flasks, each of which contained 100 mL of broth media. These flasks were then sterilized in an autoclave at standard conditions. The McFarland 0.5 methodology was used to standardize the bacterial culture and examine the compounds' antibacterial susceptibility. In addition, the medium was supplemented with composite materials from 25 to 100 µg/mL before being grown in a rotatory incubator at 37 °C for an incubation period of 16 h. Using a broth medium, the nanocomposite materials were serially diluted before being injected with a bacterial culture containing 5×10^6 CFU for a 16 h incubation period. Bacterial culture turbidity and plate count were assessed following the incubation period.

3.2.3. Bacterial Survivability with Nanomaterials

In this study, the spot-plating method was used to quantify bacterial growth. The assay utilizes the principle that a bacterial culture would have a decreasing percentage of viable cells with increasing concentrations of anti-microbial compounds. At a constant sub-MIC level of an agent, the sensitivity of a bacterial strain could be correlated with the percentage of cells that survive and form a colony. This ratio of surviving cells to the total number of plated cells is the plating efficiency. The key to a successful spot-plating assay is to spot-plate the same number of cells onto each spot. By comparing the surviving cells on the material containing plates to the control, the sensitivity of the cells to the nanocomposite concentration by plating efficiency percentage can be determined.

Determination of the optimal sub-MIC level of nanomaterials: The chosen sub-MIC level nanomaterials should be chosen with care. The concentration should be high enough to inhibit growth, but not to inhibit all growth. This optimal sub-MIC concentration is usually just below the MIC of the nanomaterials. For example, the MIC of nanomaterials and *E. coli* is 50 µg/mL. The optimal sub-MIC of nanomaterials for the spot plate assay was found to be 40 µg/mL. Media preparation for bacterial growth with and without nanomaterials should be made. The bacterial stocks should be grown overnight at their optimal growth conditions and maintained according to the McFarland standard, culture optimum incubation at 37 °C, and shaking at 250 rpm. After overnight growth, the culture should be diluted to a concentration that would allow countable isolated colonies on the final spots of the LB plates. We recommend dilutions of 1×10^4 CFU/mL. A small volume (10 µL) of the working stock of bacterial culture should be plated up to six times on each plate. The number of colonies in each spot should be counted and tallied. The number of colonies on the antibiotic plates and the control plates are used to calculate.

Nanomaterials stock preparation: The amount of the required nanomaterial concentration was weighed by analytical balance and then transferred and dissolved with the required amount of sterile dH2O, before being filtered through a 0.20 µm filter into a sterile 10 mL falcon tube.

Dilution of bacterial culture to a working concentration: The bacterial culture turbidity and availability of bacteria count was determined by spectrophotometer at OD600. The obtained value was multiplying the OD600 of the bacterial culture by the appropriate conversion ratio of OD600 to CFU/mL (if OD600 = 8×10^8 CFU/mL). Further specific required dilution of the working concentration was also determined.

Spot plating: First of all, the media plates were prepared and supplemented with and without nanomaterials of interest at the desired concentrations of 0 to 100 µg. The culture tubes were inoculated and incubated overnight at 37 °C. The next day, the optical density of the diluted cultures was determined and converted into CFU/mL using the conversion factor of the strain, if known (1 OD at 600 equals to 8×10^8 cells/mL). Further, the cultures (100 µL of culture into 900 µL of fresh growth media) were serially diluted to obtain a 1×10^4 CFU/mL culture. An aseptic pipette was used to separate the spots onto a plate using 10 µL of 1×10^4 CFU/mL culture. After incubation, the isolated colonies per spot were counted. The number of colonies on each spot of the nanomaterial's plates were also counted.

3.3. Imaging of Bacterial Growth Inhibition by SEM

In this study, both test bacterial strain fresh cultures were taken according to the McFarland standard and centrifuged and treated with the nanocomposite material, according to the MIC concentration, for 16 h incubation at 37 °C before being centrifuged at 8000 rpm for 10 min. The bacterial pellets were selected and washed multiple times with phosphate saline buffer and treated with 2% glutaraldehyde solution and the bacterial sample was placed at 4 °C for proper fixing. After fixing, the bacterial sample was washed with double distilled water, then further washed with 10 to 100 % ethanol in increasing order, and the obtained bacterial culture was then mounted on the stab of SEM and analyzed.

4. Conclusions

In this study, the nanocomposite material GN@CR-Cu was successfully synthesized and coated on contact surfaces for studying its antimicrobial activity. The antimicrobial coatings of GN@CR-Cu were prepared using the chemical synthesis methodology and were further characterized using electron microscopy and X-ray photon spectroscopy. GN@CR-Cu showed excellent antimicrobial effects against *E. coli* and *P aeruginosa* bacterial isolates. The nanocomposite showed antimicrobial activity, most likely by contact-killing mechanisms, which was suggested by zone inhibition and scanning electron microscopy. The materials showed maximum antibacterial activity against *E. coli* (24 ± 0.50 mm) followed by *P. aeruginosa* (18 ± 0.25 mm) at 25 µg/mL spot inoculation on the solid media plate, and a similar trend was observed in the minimum inhibition concentration (80 µg/mL) and bactericidal concentration (160 µg/mL) in liquid media. According to this proof-of-concept study, GN@CR-Cu can function as a potent future anti-microbial nanomaterial for the prevention of nosocomial infection, if coated on medical devices or food preparation instruments.

Supplementary Materials: The following supporting information can be downloaded at: https://www.mdpi.com/article/10.3390/molecules28062814/s1, Figure S1: FESEM image of (a) CR and (b) Cu nanoparticles, Figure S2: XPS survey scan of GN and Cu, Figure S3: Water contact angle measurement of GN@CR-Cu composite.

Author Contributions: Conceptualization, M.O., M.O.A. and A.M.; data curation, M.S.A. and A.M.; formal analysis, M.O.A. and M.S.A.; investigation M.S.A. and A.M.; methodology, M.O. and M.O.A.; supervision, A.M.; validation, M.O. and M.O.A.; writing—original draft, M.O., M.O.A. and M.S.A.; writing—review and editing, M.S.A. and A.M. All authors have read and agreed to the published version of the manuscript.

Funding: This research received no external funding.

Institutional Review Board Statement: Not applicable.

Informed Consent Statement: Not applicable.

Data Availability Statement: The data presented in this study are available on request from the corresponding author.

Conflicts of Interest: The authors declare no conflict of interest.

References

1. Carrascosa, C.; Raheem, D.; Ramos, F.; Saraiva, A.; Raposo, A. Microbial Biofilms in the Food Industry-A Comprehensive Review. *Int. J. Environ. Res. Public Health* **2021**, *18*, 2014. [CrossRef] [PubMed]
2. Lan, X.; Zhang, H.; Qi, H.; Liu, S.; Zhang, X.; Zhang, L. Custom-design of triblock protein as versatile antibacterial and biocompatible coating. *Chem. Eng. J.* **2023**, *454*, 140185. [CrossRef]
3. Birkett, M.; Dover, L.; Lukose, C.C.; Zia, A.W.; Tambuwala, M.M.; Serrano-Aroca, A. Recent Advances in Metal-Based Antimicrobial Coatings for High-Touch Surfaces. *Int. J. Mol. Sci.* **2022**, *23*, 1162. [CrossRef]
4. Yousefi, M.; Dadashpour, M.; Hejazi, M.; Hasanzadeh, M.; Behnam, B.; de la Guardia, M.; Shadjou, N.; Mokhtarzadeh, A. Antibacterial activity of graphene oxide as a new weapon nanomaterial to combat multidrug-resistance bacteria. *Mater. Sci. Eng. C* **2017**, *74*, 568–581. [CrossRef] [PubMed]
5. Sawangphruk, M.; Srimuk, P.; Chiochan, P.; Sangsri, T.; Siwayaprahm, P. Synthesis and antifungal activity of reduced graphene oxide nanosheets. *Carbon* **2012**, *50*, 5156–5161. [CrossRef]
6. Ebrahimi, M.; Asadi, M.; Akhavan, O. Graphene-based Nanomaterials in Fighting the Most Challenging Viruses and Immunogenic Disorders. *ACS Biomater. C Sci. Eng.* **2022**, *8*, 54–81. [CrossRef]
7. Mohammed, H.; Kumar, A.; Bekyarova, E.; Al-Hadeethi, Y.; Zhang, X.; Chen, M.; Ansari, M.S.; Cochis, A.; Rimondini, L. Antimicrobial mechanisms and effectiveness of graphene and graphene-functionalized biomaterials. A scope review. *Front. Bioeng. Biotechnol.* **2020**, *8*, 465. [CrossRef]
8. Roy, S.; Sarkhel, S.; Bisht, D.; Hanumantharao, S.N.; Rao, S.; Jaiswal, A. Antimicrobial mechanisms of biomaterials: From macro to nano. *Biomater. Sci.* **2022**, *10*, 4392–4423. [CrossRef]
9. Kumar, P.; Huo, P.; Zhang, R.; Liu, B. Antibacterial properties of graphene-based nanomaterials. *Nanomaterials* **2019**, *9*, 737. [CrossRef]
10. Akhavan, O.; Ghaderi, E.; Esfandiar, A. Wrapping bacteria by graphene nanosheets for isolation from environment, reactivation by sonication, and inactivation by near-infrared irradiation. *Phys. Chem. B* **2011**, *115*, 6279–6288. [CrossRef]
11. Pham, V.T.H.; Truong, V.K.; Quinn, M.D.J.; Notley, S.M.; Guo, Y.; Baulin, V.A.; Kobaisi, M.A.; Crawford, R.J.; Ivanova, E.P. Graphene Induces Formation of Pores That Kill Spherical and Rod-Shaped Bacteria. *ACS Nano* **2015**, *9*, 8458–8467. [CrossRef]
12. Jeong, J.; Kim, J.Y.; Yoon, J. The Role of Reactive Oxygen Species in the Electrochemical Inactivation of Microorganisms. *Environ. Sci. Technol.* **2006**, *40*, 6117–6122. [CrossRef] [PubMed]

13. Azam, A.; Ahmed, A.; Oves, M.; Khan, M.; Memic, A. Size-dependent antimicrobial properties of CuO nanoparticles against Gram-positive and -negative bacterial strains. *Int. J. Nanomed.* **2012**, *7*, 3527–3535. [CrossRef]
14. Akhavan, O.; Ghaderi, E. Cu and CuO nanoparticles immobilized by silica thin films as antibacterial materials and photocatalysts. *Surf. Coat. Technol.* **2010**, *205*, 219–223. [CrossRef]
15. Azam, A.; Ahmed, A.; Oves, M.; Khan, M.S.; Habib, S.S.; Memic, A. Antimicrobial activity of metal oxide nanoparticles against Gram-positive and Gram-negative bacteria: A comparative study. *Int. J. Nanomed.* **2012**, *7*, 6003–6009. [CrossRef] [PubMed]
16. Hasan, A.; Morshed, M.; Memic, A.; Hassan, S.; Webster, T.J.; Marei, H.E.-S. Nanoparticles in tissue engineering: Applications, challenges and prospects. *Int. J. Nanomed.* **2018**, *13*, 5637–5655. [CrossRef]
17. Salah, N.; Habib, S.; Khan, Z.H.; Memic, A.; Azam, A.; Alarfaj, E.; Zahed, N.; Al-Hamedi, S. High-energy ball milling technique for ZnO nanoparticles as antibacterial material. *Int. J. Nanomed.* **2011**, *6*, 863–869. [CrossRef]
18. Al-Amri, S.; Ansari, M.S.; Rafique, S.; Aldhahri, M.; Rahimuddin, S.; Azam, A.; Memic, A. Ni Doped CuO Nanoparticles: Structural and Optical Characterizations. *Curr. Nanosci.* **2015**, *11*, 191–197. [CrossRef]
19. Abdullah, T.; Qurban, R.O.; Bolarinwa, S.O.; Mirza, A.A.; Pasovic, M.; Memic, A. 3D Printing of Metal/Metal Oxide Incorporated Thermoplastic Nanocomposites With Antimicrobial Properties. *Front. Bioeng. Biotechnol.* **2020**, *8*, 568186. [CrossRef]
20. Salah, N.; Alfawzan, A.M.; Allafi, W.; Baghdadi, N.; Saeed, A.; Alshahrie, A.; Al-Shawafi, W.M.; Memic, A. Size-controlled, single-crystal CuO nanosheets and the resulting polyethylene–carbon nanotube nanocomposite as antimicrobial materials. *Polym. Bull.* **2021**, *78*, 261–281. [CrossRef]
21. Mathews, S.; Hans, M.; Mücklich, F.; Solioz, M. Contact Killing of Bacteria on Copper Is Suppressed if Bacterial-Metal Contact Is Prevented and Is Induced on Iron by Copper Ions. *Appl. Environ. Microbiol.* **2013**, *79*, 2605–2611. [CrossRef]
22. Parra, A.; Toro, M.; Jacob, R.; Navarrete, P.; Troncoso, M.; Figueroa, G.; Reyes-Jara, A. Antimicrobial effect of copper surfaces on bacteria isolated from poultry meat. *Braz. J. Microbiol.* **2018**, *49*, 113–118. [CrossRef]
23. Govind, V.; Bharadwaj, S.; Ganesh, M.R.S.; Vishnu, J.; Shankar, K.V.; Shankar, B.; Rajesh, R. Antiviral properties of copper and its alloys to inactivate COVID-19 virus: A review. *BioMetals* **2021**, *34*, 1217–1235. [CrossRef]
24. Bleichert, P.; Santo, C.E.; Hanczaruk, M.; Meyer, M.; Grass, G. Inactivation of bacterial and viral biothreat agents on metallic copper surfaces. *BioMetals* **2014**, *27*, 1179–1189. [CrossRef]
25. Noyce, J.O.; Michels, H.; Keevil, C.W. Use of Copper Cast Alloys To Control Escherichia coli O157 Cross-Contamination during Food Processing. *Appl. Environ. Microbiol.* **2006**, *72*, 4239–4244. [CrossRef]
26. Wilks, S.A.; Michels, H.; Keevil, C.W. The survival of Escherichia coli O157 on a range of metal surfaces. *Int. J. Food Microbiol.* **2005**, *105*, 445–454. [CrossRef]
27. Weaver, L.; Michels, H.T.; Keevil, C.W. Survival of Clostridium difficile on copper and steel: Futuristic options for hospital hygiene. *J. Hosp. Infect.* **2008**, *68*, 145–151. [CrossRef]
28. Schraufstatter, E.; Bernt, H. Antibacterial Action of Curcumin and Related Compounds. *Nature* **1949**, *164*, 456. [CrossRef]
29. Nelson, K.M.; Dahlin, J.L.; Bisson, J.; Graham, J.; Pauli, G.F.; Walters, M.A. The Essential Medicinal Chemistry of Curcumin. *J. Med. Chem.* **2017**, *60*, 1620–1637. [CrossRef]
30. Adamczak, A.; Ożarowski, M.; Karpiński, T.M. Curcumin, a Natural Antimicrobial Agent with Strain-Specific Activity. *Pharmaceuticals* **2020**, *13*, 153. [CrossRef]
31. Dai, C.; Lin, J.; Li, H.; Shen, Z.; Wang, Y.; Velkov, T.; Shen, J. The Natural Product Curcumin as an Antibacterial Agent: Current Achievements and Problems. *Antioxidants* **2022**, *11*, 459. [CrossRef] [PubMed]
32. Ak, T.; Gülçin, İ. Antioxidant and radical scavenging properties of curcumin. *Chem.-Biol. Interact.* **2008**, *174*, 27–37. [CrossRef] [PubMed]
33. Oves, M.; Rauf, M.A.; Ansari, M.O.; Khan, A.A.P.; Qari, H.A.; Alajmi, M.F.; Sau, S.; Iyer, A.K. Graphene Decorated Zinc Oxide and Curcumin to Disinfect the Methicillin-Resistant Staphylococcus aureus. *Nanomaterials* **2020**, *10*, 1004. [CrossRef]
34. Dolgov, A.; Lopaev, D.; Lee, C. Characterization of carbon contamination under ion and hot atom bombardment in a tin-plasma extreme ultraviolet light source. *Appl. Surf. Sci.* **2015**, *353*, 708–713. [CrossRef]
35. Ansari, M.O.; Kumar, R.; Alshahrie, A.; Abdel-wahab, M.S.; Sajith, V.K.; Ansari, M.S.; Jilani, A.; Barakat, M.A.; Darwesh, R. CuO sputtered flexible polyaniline@graphene thin films:A recyclable photocatalyst with enhanced electrical properties. *Compos. Part B Eng.* **2019**, *175*, 107092. [CrossRef]
36. Akhavan, O.; Ghader, E.; Shirazian, S.A. Near infrared laser stimulation of human neural stem cells into neurons on graphene nanomesh semiconductors. *Colloids Surf B Biointerfaces* **2015**, *126*, 313–321. [CrossRef]
37. Akhavan, O.; Tohidi, H.; Moshfegh, A. Synthesis and electrochromic study of sol–gel cuprous oxide nanoparticles accumulated on silica thin film. *Thin Solid Film.* **2009**, *517*, 6700–6706. [CrossRef]
38. Li, J.; He, M.; Yan, J.; Liu, J.; Zhang, J.; Ma, J. Room Temperature Engineering Crystal Facet of Cu_2O for Photocatalytic Degradation of Methyl Orange. *Nanomaterials* **2022**, *12*, 1697. [CrossRef]
39. Ozdal, M.; Gurkok, S. Recent advances in nanoparticles as antibacterial agent. *ADMET DMPK* **2022**, *10*, 115–129. [CrossRef]
40. Ahmad, N.S.; Abdullah, N.; Yasin, F.M. Toxicity assessment of reduced graphene oxide and titanium dioxide nanomaterials on gram-positive and gram-negative bacteria under normal laboratory lighting condition. *Toxicol. Rep.* **2020**, *7*, 693–699. [CrossRef]
41. Culotti, A.; Packman, A.I. Pseudomonas aeruginosa Promotes Escherichia coli Biofilm Formation in Nutrient-Limited Medium. *PLoS ONE* **2014**, *9*, e107186. [CrossRef]

42. Sanni, O.; Chang, C.-Y.; Anderson, D.G.; Langer, R.; Davies, M.C.; Williams, P.M.; Williams, P.; Alexander, M.R.; Hook, A.L. Bacterial Attachment to Polymeric Materials Correlates with Molecular Flexibility and Hydrophilicity. *Adv. Healthc. Mater* **2015**, *4*, 695–701. [CrossRef]
43. Rosenhahn, A.; Schilp, S.; Kreuzerc, H.J.; Grunze, M. The role of "inert" surface chemistry in marine biofouling prevention. *Phys. Chem. Chem. Phys.* **2010**, *12*, 4275–4286. [CrossRef]
44. Hojati, S.T.; Alaghemand, S.; Hamze, F.; Babaki, F.A.; Rajab-Nia, R.; Rezvani, M.B.; Kaviani, M.; Atai, M. Antibacterial, physical and mechanical properties of flowable resin composites containing zinc oxide nanoparticles. *Dent. Mater.* **2013**, *29*, 495–505. [CrossRef]
45. Lai, M.-J.; Huang, Y.-W.; Chen, H.-C.; Tsao, L.-I.; Chien, C.-F.C.; Singh, B.; Liu, B.R. Effect of Size and Concentration of Copper Nanoparticles on the Antimicrobial Activity in Escherichia coli through Multiple Mechanisms. *Nanomaterials* **2022**, *12*, 3715. [CrossRef]
46. Hettiarachchi, S.S.; Perera, Y.; Dunuweera, S.P.; Dunuweera, A.N.; Rajapakse, S.; Rajapakse, M.G.R. Comparison of Antibacterial Activity of Nanocurcumin with Bulk Curcumin. *ACS Omega* **2022**, *7*, 46494–46500. [CrossRef]
47. Wang, L.; Hu, C.; Shao, L. The antimicrobial activity of nanoparticles: Present situation and prospects for the future. *Int. J. Nanomed.* **2017**, *12*, 1227–1249. [CrossRef]
48. Akhavan, O.; Ghaderi, E. Toxicity of Graphene and Graphene Oxide Nanowalls Against Bacteria. *ACS Nano* **2010**, *4*, 5731–5736. [CrossRef]
49. Dutta, T.; Sarkar, R.; Pakhira, B.; Ghosh, S.; Sarkar, R.; Barui, A.; Sarkar, S. ROS generation by reduced graphene oxide (rGO) induced by visible light showing antibacterial activity: Comparison with graphene oxide (GO). *RSC Adv.* **2015**, *5*, 80192–80195. [CrossRef]
50. Prasanna, V.L.; Vijayaraghavan, R. Insight into the Mechanism of Antibacterial Activity of ZnO: Surface Defects Mediated Reactive Oxygen Species Even in the Dark. *Langmuir* **2015**, *31*, 9155–9162. [CrossRef]
51. Li, S.; Zeng, T.H.; Hofmann, M.; Burcombe, E.; Wei, J.; Jiang, R.; Kong, J.; Chen, Y. Antibacterial Activity of Graphite, Graphite Oxide, Graphene Oxide, and Reduced Graphene Oxide: Membrane and Oxidative Stress. *ACS Nano* **2011**, *5*, 6971–6980. [CrossRef] [PubMed]
52. Akhavan, O.; Ghaderi, E. Escherichia coli bacteria reduce graphene oxide to bactericidal graphene in a self-limiting manner. *Carbon* **2012**, *50*, 1853–1860. [CrossRef]
53. Kumar, A.; Pandey, A.K.; Singh, S.S. Engineered ZnO and TiO$_2$ nanoparticles induce oxidative stress and DNA damage leading to reduced viability of Escherichia coli. *Free. Radic. Biol. Med.* **2011**, *51*, 1872–1881. [CrossRef] [PubMed]
54. Wang, Y.-W.; Cao, A.; Jiang, Y.; Zhang, X.; Liu, J.-H.; Liu, Y.; Wang, H. Superior Antibacterial Activity of Zinc Oxide/Graphene Oxide Composites Originating from High Zinc Concentration Localized around Bacteria. *ACS Appl. Mater. Interfaces* **2014**, *6*, 2791–2798. [CrossRef]
55. Jannesari, M.; Akhavan, O.; Hosseini, H.R.M.; Bakhshi, B. Graphene/CuO$_2$ Nanoshuttles with Controllable Release of Oxygen Nanobubbles Promoting Interruption of Bacterial Respiration. *ACS Appl. Mater. Interfaces* **2020**, *12*, 35813–35825. [CrossRef]
56. Khan, A.; Rashid, A.; Younas, R.; Chong, R. A chemical reduction approach to the synthesis of copper nanoparticles. *Int. Nano Lett.* **2015**, *6*, 21–26. [CrossRef]
57. Marcano, D.C.; Kosynkin, D.V.; Berlin, J.M.; Sinitskii, A.; Sun, Z.; Slesarev, A.; Alemany, L.B.; Lu, W.; Tour, J.M. Improved Synthesis of Graphene Oxide. *ACS Nano* **2010**, *4*, 4806–4814. [CrossRef]

Disclaimer/Publisher's Note: The statements, opinions and data contained in all publications are solely those of the individual author(s) and contributor(s) and not of MDPI and/or the editor(s). MDPI and/or the editor(s) disclaim responsibility for any injury to people or property resulting from any ideas, methods, instructions or products referred to in the content.

Article

Polysaccharides as Green Fuels for the Synthesis of MgO: Characterization and Evaluation of Antimicrobial Activities

Nayara Balaba [1], Silvia Jaerger [1], Dienifer F. L. Horsth [1,2], Julia de O. Primo [1], Jamille de S. Correa [1], Carla Bittencourt [2,*], Cristina M. Zanette [3] and Fauze J. Anaissi [1]

1. Departamento de Química, Universidade Estadual do Centro-Oeste, Guarapuava 85040-080, Brazil
2. Chimie des Interactions Plasma-Surface (ChIPS), Research Institute for Materials Science and Engineering, University of Mons, 7000 Mons, Belgium
3. Departamento de Engenharia de Alimentos, Universidade Estadual do Centro-Oeste, Guarapuava 85040-080, Brazil
* Correspondence: carla.bittencourt@umons.ac.be

Abstract: The synthesis of structured MgO is reported using feedstock starch (route I), citrus pectin (route II), and *Aloe vera* (route III) leaf, which are suitable for use as green fuels due to their abundance, low cost, and non-toxicity. The oxides formed showed high porosity and were evaluated as antimicrobial agents. The samples were characterized by energy-dispersive X-ray fluorescence (EDXRF), X-ray diffraction (XRD), Fourier-transform infrared spectroscopy (FTIR), and scanning electron microscopy (SEM). The crystalline periclase monophase of the MgO was identified for all samples. The SEM analyses show that the sample morphology depends on the organic fuel used during the synthesis. The antibacterial activity of the MgO-St (starch), MgO-CP (citrus pectin), and MgO-Av (*Aloe vera*) oxides was evaluated against pathogens *Staphylococcus aureus* (ATCC 6538P) and *Escherichia coli* (ATCC 8739). Antifungal activity was also studied against *Candida albicans* (ATCC 64548). The studies were carried out using the qualitative agar disk diffusion method and quantitative minimum inhibitory concentration (MIC) tests. The MIC of each sample showed the same inhibitory concentration of 400 µg.mL^{-1} for the studied microorganisms. The formation of inhibition zones and the MIC values in the antimicrobial analysis indicate the effective antimicrobial activity of the samples against the test microorganisms.

Keywords: eco-friendly synthesis; *Aloe vera*; starch; MIC; bacteria; antifungal

1. Introduction

Magnesium oxide (MgO) is an ionic material with a refractory capacity to withstand high temperatures [1]. It presents a well-defined crystalline structure of face-centered cubic, and, when hydrated, it converts to its hydroxide form (Mg(OH)$_2$) [2]. MgO is a functional, low-cost, environmentally safe metal oxide applied in various industrial fields, e.g., the plastics, rubber, paper, and adhesive industries; in agriculture, MgO is incorporated into animal feed and fertilizer, and it is used in refractory applications in steel production and equipment coatings [3]. MgO has also been studied for the adsorption of textile dyes, metal ions, and phosphates in the catalysis applied in ceramic materials and paints [1–7]. Several methodologies for the synthesis of MgO can be found: precipitation [2], microwave-assisted [4,8], sol–gel [9], and hydrothermal [5]. The main production method is by calcining dolomite (CaMg(CO$_3$)$_2$) and brucite (Mg(OH)$_2$); through this process, decomposition occurs at a high temperature [3]. However, the MgO obtained with this process is divergent due to the low specific surface area, irregular morphology, and grain size [3]. Reports on other methodologies for the synthesis of structured magnesium oxide are available [2,4,5] but the search for organic fuels to be used in synthesis has been considered of importance due to their ecological origin, easy access to natural polysaccharides, low cost, high chemical reactivity, high combustion power, reduction in the calcination temperature, and action as a complexing gelling agent [10].

In the present work, starch, citrus pectin, and *Aloe vera* leaf were used as fuels to obtain structured MgO by the combustion method. In synthesis route I, starch extracted from cassava (*Manihot esculenta*), a high-energy tuber and a low-cost, biodegradable polysaccharide from renewable raw material sources, consisting of amylose and amylopectin molecules that are composed of D-glucose units, was used [11]. In route II, citrus pectin, a polysaccharide derived from the peels of citrus fruits (lemon, orange, etc.) used in the food and pharmaceutical industries, due to its high gelling capacity, was used [12]. It consists of α-D-galacturonic acid units joined by glycosidic bonds (α-1,4) and esterified methyl carboxyl groups [12]. Finally, *Aloe vera* (*Aloe Barbadensis Miller*) was used in route III, a succulent perennial of the family Liliaceae, used for its pharmacological properties [13]. According to Hamman (2008), the *Aloe vera* leaves have three structural components: the cell walls, the degenerated organelles, and the viscous liquid contained within the cells. These components present many compounds, such as proteins, lipids, amino acids, vitamins, enzymes, inorganic compounds, small organic compounds, and polysaccharides. Among the polysaccharides, one can find mainly mannose, cellulose, and pectic polysaccharides, whereas the skin of the leaf contains, in addition, significant quantities of xylose-containing polysaccharides [13,14]. Describing a reaction mechanism for the synthesis of materials with *Aloe vera* is complex due to the different compounds present in the plant extract [15]. However, a sustainable reaction mechanism for the formation of material with organic fuel is the interaction of metal ions that bind with biomolecules through functional groups and π electrons by ionic bonds or van der Waals forces; this depends on the concentration of plant extracts [15]. The synthesized MgO samples were characterized, and their biological activities were studied.

Antimicrobial resistance (AMR) is a global threat to human health, causing thousands of deaths annually [16]. One of the leading causes of drug-resistant pathogens is the excessive exposure of microorganisms to antibiotics as a treatment against infection [17,18]. Another problem caused by microorganisms is biofilms, which are aggregates of cells embedded in a self-producing matrix of extracellular polymeric substances (EPS), which adhere to each other and/or to a surface and have greater potential to survive in adverse conditions and end up generating resistance to antibiotic treatment and the host's immune system [19]. Bacteria and diverse fungi show greater susceptibility to antibiotic resistance due to some strains' evolutionary and adaptive conditions [17,20]. For example, the bacterium *Staphylococcus aureus*, responsible for infections such as postoperative wounds and prosthetic infections related to endotracheal tubes and other biomaterials, has been reported to cause nosocomial infections and AMR to several drugs, such as penicillin, methicillin, quinolone, and vancomycin [17,21]. A fungus that has become resistant to antibiotics is *Candida albicans*, an opportunistic pathogen, generally harmless to human beings, which can be found on the surface of humid mucous, such as in the intestine, vagina, and oral cavity [17]. The fungus *C. albicans* shows drug resistance to amphotericin B and azoles [22], two known antifungal agents. However, colonization, infectious aggravation, and damage to the structure of cells may occur with the weakening of the host's immune system [17]. Therefore, it is necessary to develop alternative strategies to minimize the problem of AMR, such as materials that inhibit the growth of microorganisms, preventing contagion with infectious diseases [23]. In this context, some papers present antimicrobial studies, as Al-Shammar et al. (2021), which uses zein nanoparticles loaded with transition metal ions to control three *Candida* species [24], and metal oxide particles have presented a wide range of biological applications, such as drug and gene delivery and cell, tissue, and diagnostics engineering, as well as limiting the growth of microorganisms [4]. Reports on structured magnesium oxide (MgO) describe its efficiency in inhibiting the growth of food and aquatic pathogen colonies and controlling the proliferation of bacteria and fungi, attracting significant interest due to its non-toxic nature [4,5,25]. The common mechanism of the antibacterial activity of MgO is due to the oxygen vacancies, leading to the higher production of reactive oxygen species (ROS) (OH$^-$, O^{2-}, and H$_2$O$_2$) on the surfaces of the particles when oxidative stress occurs on the bacterial cell wall, leading to cell death [4]. Another possible mechanism is when Mg^{2+} ions are released from MgO with irregular morphology and rough edges, coming into contact with the cell membrane of the microbe [4,8]. The negatively charged cell membrane and Mg^{2+} attract each other and the Mg penetrates the cell membrane and damages it, leading to cell death [8]. Therefore, in this study, three natural polysaccharides were used as precursors to the combustion reaction to obtain structured MgO, aiming at its application as an antibacterial and antifungal agent.

2. Results

2.1. Characterization of Synthesized MgO Particles

According to the semiquantitative analysis (EDXRF) (Table 1), the sample with the highest elemental magnesium percentage was MgO-St, with 86.6% atomic, while MgO-Av presented 64.1%, and MgO-CP 76.2% (Figure S1). This difference can be associated with the composition of the organic additives: starch contained ~6.6% atomic Mg ions, citrus pectin presented ~1.7% of magnesium, and *Aloe vera* gel contained ~0.1% (Figure S2). The composition of the organic additives varied according to the region from which they were harvested [26]. The samples' composition variation indicates that the organic fuel used in the synthesis influences the amount of metal ions in the material obtained, interfering with the physicochemical properties and affecting the performance against bacteria and fungi.

Table 1. Elemental chemical compositions of the synthesized MgO samples and precursors used in the synthesis in element percent (% element) by EDXRF.

Sample	Elements (%)								
	Mg	Ca	Zn	Cu	K	Al	S	P	Others
Mg(NO$_3$)$_2$·6H$_2$O	98.0	0.8	-	0.1	-	-	1.0	-	0.1
Starch	6.6	26.0	4.0	4.4	22.5	15.2	11.6	4.9	4.8
MgO-St	86.6	2.7	0.7	0.4	1.0	6.3	-	1.7	0.6
Aloe vera	0.1	20.9	1.0	3.4	0.5	-	1.2	-	72.9
MgO-Av	64.1	8.8	0.4	0.7	0.4	-	1.1	0.4	24.1
Citric pectin	1.7	21.8	2.2	2.1	52.6	7.7	4.9	1.4	5.6
MgO-CP	76.2	8.1	1.6	0.5	1.3	7.3	2.6	0.5	1.9

Figure 1 shows the X-ray diffractograms of magnesium oxide samples obtained from starch (Figure 1a), citrus pectin (Figure 1b), and *Aloe Vera* (Figure 1c). In all samples, peaks (111), (200), (220) (113), and (222) of the periclase crystalline phase, the typical phase of the face-centered cubic MgO (COD: 9000505), were observed [27,28].

A few differences between the diffractograms were observed when using different fuels. The samples showed different peak intensities, mainly peak (200) at 42.6° and peak (220) at 61.8°, and MgO-St was found to have a greater width at the average height compared to the peaks of the MgO-CP and MgO-Av diffractograms, which would suggest a smaller crystallite size [28].

The FT-IR spectra of the MgO samples are shown in Figure 2. The IR spectra exhibit a narrow and intense band at 3700 and 3445 cm^{-1}, attributed to the vibrational stretching of the free -OH ions of Mg(OH)$_2$, generated by the hydration of MgO [29]. The γ(OH) region shows two bands at 1488 cm^{-1} for the MgO-St sample (Figure 2a) and 1638 cm^{-1}, which correspond to the O–H bending mode, which is characterized by the bending vibration of the -OH group of the physiosorbed water molecules [29,30]. The broad bands at 1427 and 1383 cm^{-1} for MgO-St, MgO-CP, and MgO-Av (Figure 2b,c) are assigned to the asymmetrical and symmetrical stretching vibrations of CO$_2$ species chemisorbed onto the surface of MgO [31,32]. The low-intensity bands at 1110 and 1061 cm^{-1} observed for MgO-CP and Mg-Av can be associated with the presence of H ion species as defects of octahedral symmetry, characteristic of magnesium oxides calcined between 700 and 800 °C [33,34].

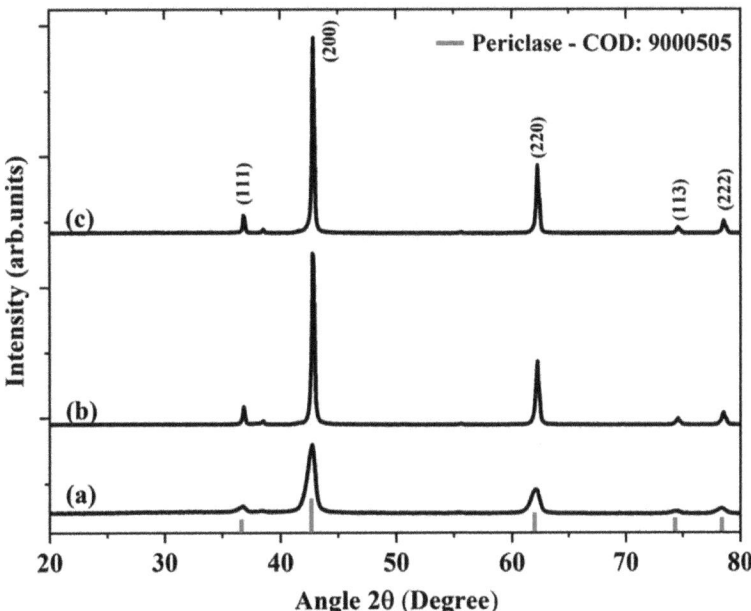

Figure 1. XRD patterns of samples MgO-St (**a**), MgO-CP (**b**), and MgO-Av (**c**).

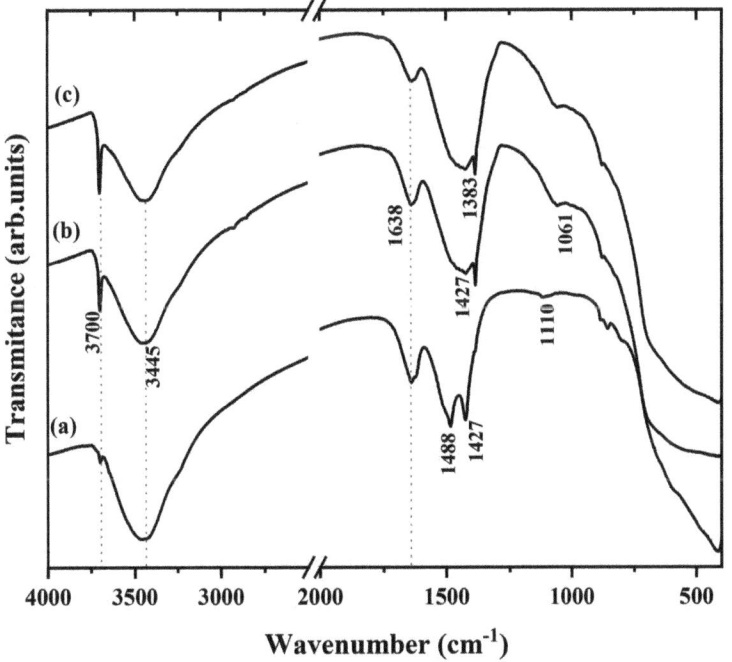

Figure 2. FTIR spectra for the samples (**a**) MgO-St, (**b**) MgO-CP, and (**c**) MgO-Av.

Scanning electron microscopy was used to study the morphology of the synthesized samples (Figure 3). The morphology of the MgO-St sample (Figure 3a) displayed a structure composed of non-uniform pores and hole voids, with spongy characteristics in the material, with an average particle size of approximately 0.99 µm (Figure 3b). The holes may be due to the large number of gases released during the combustion of the reagents and the starch used in the synthesis. These voids and pores provide a large surface area, supporting antimicrobial activity [35]. Figure 3c shows SEM images of the MgO-CP sample, which had an irregular morphology, forming small pseudo-spheres and hole voids [20], showing larger particles than the MgO-St sample, with an average size of 1.14 µm (Figure 3d). Finally, the morphology of the MgO-Av sample (Figure 3e) was irregular and composed of periodic sheets, and the sample consisted of small particles with an average size of 0.85 µm (Figure 3f). The different morphologies observed in the samples are attributed to the different polysaccharides used in each type of synthesis, which will affect the antimicrobial properties of each material [35–37].

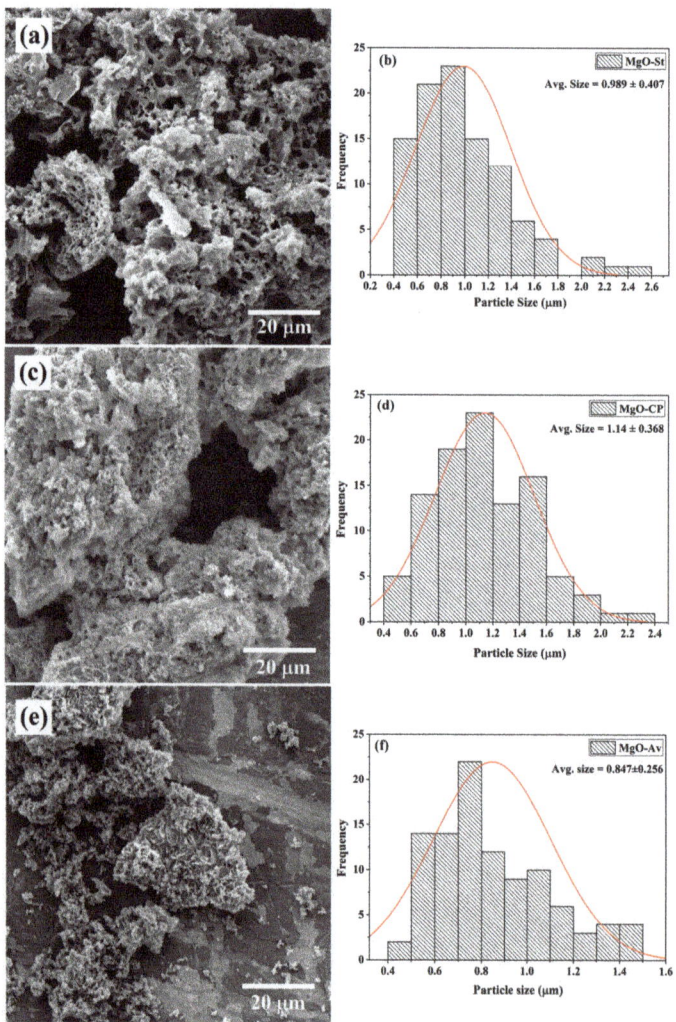

Figure 3. SEM images of synthesized MgO using different natural additives as fuel sources and average particle sizes of the samples: (**a**,**b**) MgO-St, (**c**,**d**) MgO-CP, and (**e**,**f**) MgO-Av.

2.2. Evaluation of Antimicrobial Activity

The MgO samples were tested against *C. albicans* (ATCC 64548) using the disk diffusion test; see Figure 4. The clear zones around the specimens indicate the inhibition of the fungal growth. This halo of inhibition (in millimeters) was used to quantify the antifungal activity obtained, and their averages are presented in Table 2. The sample with the highest antifungal activity is the one with the largest halo around the disk; the diffusion of the antimicrobial agent leads to the formation of a zone of inhibition of bacterial growth, whose diameter is proportional to the inhibition [38]. The MgO-Av sample shows weak antifungal activity. In contrast, the MgO-St sample exhibits an inhibition zone with a diameter of 3.0 mm on average. The MgO-St samples have the largest surface area. Several studies have correlated a large surface area with good antibacterial and antifungal activity performance [20], suggesting that the superior inhibition observed for the MgO-St sample is associated with its surface characteristics and morphology.

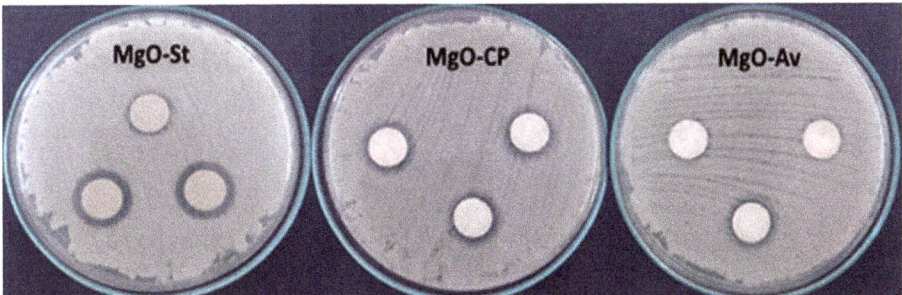

Figure 4. Photographs of halos of inhibition formed in tests against *C. Albicans* using the oxide samples MgO-St, MgO-CP, and MgO-Av. The oxide MgO-St showed the largest halo of inhibition.

Table 2. Average mean zones of inhibition (in mm) produced by synthesized MgO samples on the test organisms.

Microorganism	Disk Diffusion Method		
	MgO-St	MgO-CP	MgO-Av
Staphylococcus aureus	2.86 ± 0.23 [a]	2.43 ± 0.51 [b]	2.20 ± 0.34 [b]
Escherichia coli	0	0	0
Candida albicans	3.0 ± 0 [a]	1.36 ± 0.11 [b]	0.16 ± 0.28 [c]

[a,b,c] Results are represented as the mean ± standard deviation. Different lowercase letters in the same row indicate significant differences at $p \leq 0.05$ by Tukey's test.

MgO is reported to have antibacterial properties, including excellent antibacterial activity against *K. pneumonia* bacteria [35,39]. In this context, we investigated its antibacterial activity against *S. aureus* and *E. coli*; the MgO-synthesized samples exhibited higher inhibition than the control (Figure 5).

The MgO samples were tested against pathogenic Gram-positive *S. aureus* and Gram-positive *E. coli* bacteria using the disk diffusion methodology. Table 2 shows the means and standard deviations of the inhibition zone; the halo or zone of inhibition can be observed in Figure 5 by the darker zones around the disks. It was found that all the MgO samples had antibacterial properties against *S. aureus*. The MgO-St sample had statistically higher activity against *S. aureus* and *C. albicans*, with a larger inhibition diameter, which might explain its larger surface area compared to the other samples (MgO-Av and MgO-CP).

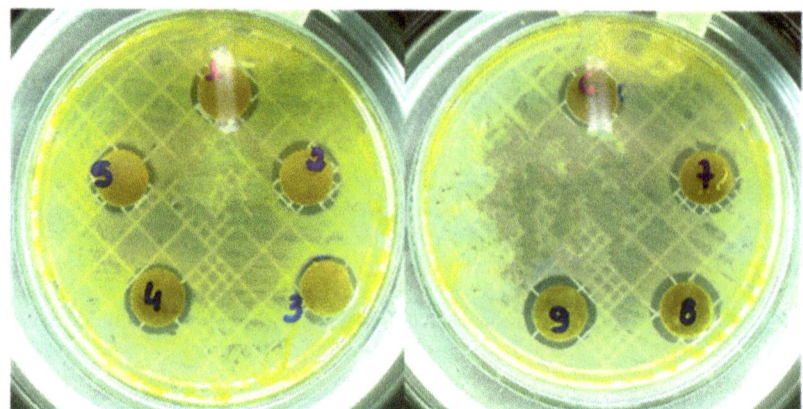

Figure 5. Photographs of halos of inhibition formed in tests against *Staphylococcus aureus*. *MgO-CP (labeled as 1,2,3); MgO-Av (labelled as 4,5,6); MgO-St (labeled as 7,8,9).

It has been reported that the antibacterial activity increases with the decreasing size of nanoparticles due to the bacteria MgO surface interaction, which depends on the surface area available [20]. Against *S. aureus*, the oxides synthesized with citric pectin and *Aloe vera* fuels were statistically equal; however, against the *C. albicans* fungi, the oxide MgO-CP showed significantly higher activity than the MgO-Av sample. The MgO samples were ineffective against *E. coli*, which can be associated with the Gram-negative bacterial composition of a thin peptidoglycan cell wall and an outer membrane containing lipopolysaccharide, leading to higher resistance than for Gram-positive bacteria such as *S. aureus* [20]. Some studies, such as Umaralikhan and Jaffar, also used *Aloe vera* leaves as a fuel source for MgO synthesis and tested them against *S. aureus* and *E. coli* as an antibacterial material in the disk diffusion method; they obtained considerable inhibition success against both bacteria [25]. The diffusion of agar disks is a qualitative assay used to test bioactivity in a sample. However, in this method, the effect of antimicrobial activity is not accurately estimated [40]. Therefore, the synthesized oxides were tested by the minimum inhibitory concentration (MIC) method to quantify the inhibitory concentration for each strain. The values for the MIC provide a quantitative evaluation of the antimicrobial action of the samples against *E. coli* and *S. aureus* pathogens. The MIC is defined as the lowest concentration ($\mu g \cdot mL^{-1}$) of an antimicrobial agent, which, under strictly in vitro conditions, completely prevents the growth of the test strain of an organism (EUCAST, 1998) [41]. The MIC of each sample showed the same inhibitory concentration of 400 $\mu g \cdot mL^{-1}$ for both bacteria *S. aureus* and *E. coli*, and also showed a minimum inhibitory concentration for *C. albicans* of 400 $\mu g \cdot mL^{-1}$. The broth microdilution method allowed the quantification of the minimum inhibitory concentration; however, the same did not occur in the agar diffusion disk tests. It can also be observed that, unlike the broth microdilution method, the three samples did not present inhibitory activity in the disc diffusion tests against *E. coli* bacteria. The absence of the inhibition zone does not necessarily indicate that the sample is inactive against the tested strain but that the diffusion may be incomplete; this occurs especially in compounds that diffuse more slowly in a solid culture medium or due to the lipophilic characteristics or/and the chemical nature of the isolated substances [42,43].

The antibacterial efficacy of MgO samples has been explained by a set of factors that inhibit bacterial growth, such as the production of reactive oxygen species (ROS) due to the presence of Mg^{2+}, the interaction between MgO particles and the membrane cell wall, and the penetration of individual particles into cells [39]. The antibacterial properties may also be associated with the correlation between the surface area and volume of oxide particles, which form more active oxygen species outside the bacterial cell, destroying the bacteria's cell membranes [8,20].

3. Discussion

The samples synthesized in this work showed differences in their structural, morphological, and antimicrobial properties, associated with the different polysaccharide sources used in each type of synthesis. The three MgO samples presented the same crystalline structure of the face-centered cubic type, characteristic of magnesium oxides; however, the MgO-St had a smaller crystallite size compared to MgO-CP and MgO-Av. The same structure was reported by Umaralikhan and Jaffar [25] and El-Shaer et al. [44]. El-Shaer et al. synthesized MgO using the conventional sol–gel method and annealed it at different temperatures for 5 h; the samples calcined above 500 °C showed similar antimicrobial activity results as in this work. According to these authors, the calcination temperature affects the number of surface defects and therefore the antimicrobial properties [44], justifying the antimicrobial results presented for the MgO-St, MgO-Av, and MgO-CP samples in the MIC test.

Some additional FTIR bands were found in the three synthesized oxides that resembled the FTIR bands found by Karthik et al. (2017) [4] at 3445, 1383, and 1638 nm. According to the authors, MgO was obtained with the same crystallographic phase by microwave-assisted green synthesis and calcining at 400 °C for 2 h. The results of the disk diffusion test for MgO obtained by Karthik et al. [4] were superior to the results reported in this work. As previously mentioned, the disk diffusion test is a qualitative method as it is limited by the mobility of the oxide in the disks and does not present the inhibition results as a concentration value. However, the disk diffusion test is a qualitative method as it is limited by the mobility of the oxide in the disk, and the inhibition results are not directly connected to the oxide concentration. Nevertheless, the MgO-St, MgO-CP, and MgO-Av samples in the MIC test showed higher inhibition against *E. coli*, *S. aureus*, and *C. albicans* than the samples reported in [45].

The sample morphology varied with the fuel used. MgO-St presented a sponge-like morphology, while MgO-CP and MgO-Av presented pseudo-spheres and lamellar morphologies, respectively. According to the literature [2,7], magnesium oxide presents several possible morphologies, depending on the synthesis method employed or the fuel used in the synthesis. In the work of Bindhu et al. [20], MgO nanoparticles were obtained via a chemical precipitation reaction using magnesium nitrate and sodium hydroxide calcinated at 400 °C for 5 h. The particles were almost spherical in shape, with smooth surfaces, and this morphology is very similar to that of MgO-CP. The antimicrobial activity studied by Bindhu et al. [20] was tested against *S. aureus* using the disk diffusion method, and the zones of inhibition were similar to those found in this work. However, the MgO nanoparticles synthesized by Bindhu et al. [20] were not tested against *E. coli*, but rather against the Gram-negative pathogen *Pseudomonas aeruginosa*.

The average particle diameter sizes for the synthesized samples ranged from 0.85 to 1.14 µm. For the MgO-St oxide, the average particle diameter size was 0.99 µm; this sample was the most porous, justifying its higher antimicrobial activity [36]. According to Ananda et al. [35], the presence of pores provides a large surface area, which supports high antimicrobial activity. In addition, this high activity can also be associated with the higher percentage of Mg ions in this sample, as a possible mechanism of antimicrobial action is the release of Mg^{2+} ions, as cited earlier.

Both the antimicrobial activity and the crystalline structures of our samples were very similar to those found in the literature; however, in this work, we used the lowest calcination time, only one hour, while other studies have reported calcination times greater than 5 h using traditional synthesis methods [20,44,45]. However, differences in the samples' morphology can be observed, which may explain the differences in the antimicrobial activity, associated with the small particle size and high porosity, favoring a larger contact area.

4. Materials and Methods

4.1. Materials

Three different fuels were used to synthesize magnesium oxide: cassava starch (route I), citrus pectin (route II), and *Aloe vera* (route III). The cassava roots used for the extraction of starch were harvested in Palmital, Paraná, Brazil, and the *Aloe vera* leaves were harvested in São José, Paraná, Brazil. The other materials used were magnesium nitrate hexahydrate ($Mg(NO_3)_2 \cdot 6H_2O$, 98%, Dynamics) and commercial citrus pectin (Dynamics). The analytical reagents were of high purity, and all the solutions were prepared with deionized water.

The same calcination parameters were used in the three synthesis routes. First, the suspensions were calcined in a muffle furnace at a temperature of 750 °C with a heating ramp of 10 °C min^{-1} for 60 min. The products obtained were pulverized, sieved on a 250 mm (60 mesh) sieve, and stored in a suitable container for characterization and application. The MgO samples obtained by the three different synthetic routes were labeled MgO-St (Route I), MgO-CP (Route II), and MgO-Av (Route III).

4.2. Synthesis Using Cassava Starch (Route I)

To obtain MgO with the starch additive, the methodology used was adapted from Primo et al. [46]. Initially, 500 g of natural starch from manioc was extracted in 2500 mL of deionized water under mechanical agitation for 3 h. Then, the colloidal suspension was sieved and used as the base solution. The oxide MgO-St was synthesized from starch (300 g) and magnesium nitrate (64 g) and agitated for 20 min.

4.3. Synthesis Using Citrus Pectin (Route II)

Using citrus pectin as an organic and energetic precursor, MgO was synthesized using the gelation method, adapted from Dalpasquale et al. [12]. In the first step, 1500 mL of deionized water was heated to 80 °C, and 15 g of citrus pectin was added under constant agitation (800 rpm) until it was solubilized and a colloidal suspension formed. Magnesium nitrate salt (32 g) was added to the citrus pectin colloidal suspension, remaining under agitation and temperature control for 3 h.

4.4. Synthesis Using Aloe vera Barbadensis Miller (Route III)

To obtain the magnesium oxide prepared using the gel extracted from *Aloe vera* leaves, the synthetic route was adapted from Primo et al. [10]. First, the gel of the leaves was removed and processed in a Britania B1000 blender (power: 1200 W). The gel was sieved and then kept under refrigeration (2 °C). Thus, an *Aloe vera* gel broth extract with a concentration of 90% was prepared with deionized water, and the volume was 100 mL. Subsequently, magnesium nitrate (24 g) was dissolved in the aloe extract solution under constant magnetic stirring (60 min).

Figure 6 represents the general scheme of synthesis using organic fuels for the synthesis of MgO. Metal ion complexation occurs by coordinating with the functional groups of the organic molecule, and, after calcination, there is not only the formation of the magnesium oxide but also the release of H_2O vapor and N_2 and CO_2 gas [47].

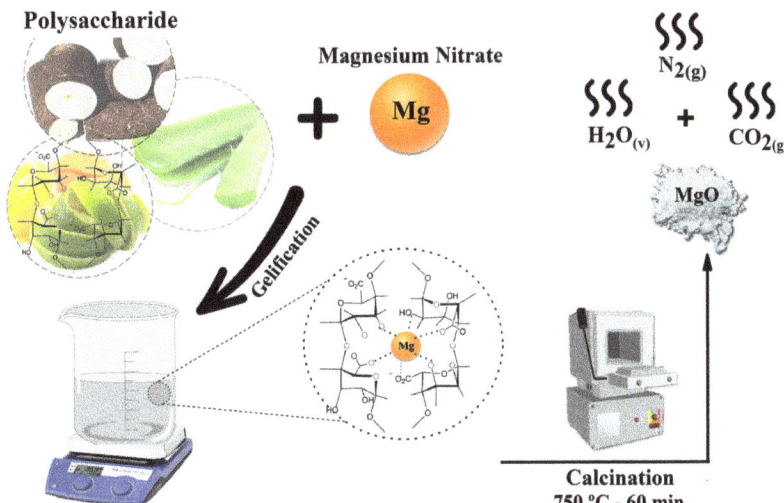

Figure 6. General schematic diagram of synthesis using polysaccharides as gelling complexing fuels.

4.5. Characterization Techniques

The elemental composition of the structured MgO was evaluated using an energy-dispersive X-ray spectrometer (EDX) (Shimadzu, Kyoto, Japan), model EDX-7000, containing an Rh tube, operating at 50 and 15 W. The crystalline structure and phase purity were characterized by X-ray diffractometry (XRD-D2 Phaser; Bruker, Billerica, MA, USA), with a copper cathode (λ = 1.5418 Å), 30 kV potential, 10 mA current, ranging between 10° and 80° (2θ) and with 0.2°/s increments. We used Phase Identification from Powder Diffraction, Version 3.4.2, with access to the Crystallography Open Database (COD). Fourier transform infrared spectroscopy was performed in a Perkin Elmer Frontier device (Waltham, MA, USA). The samples were prepared in such a way as to obtain a pellet, which consisted of a mixture of a transparent matrix to which the sample was added; we used potassium bromide (KBr) that was previously dried and kept in an oven at 100 °C for 72 h until the time of maceration with the samples. We macerated 250 mg of each oxide to 25 mg of KBr until homogenization. The ground and homogenized mixture was deposited in a steel mold (Sigma-Aldrich Z506699-1EA, St. Louis, MI, USA) and subjected to pressure of approximately 3.0 kg.cm^{-2} in a hydraulic press. With the procedure, 13-mm-diameter pellets were obtained for the formation of absorbance or transmittance spectra with a wider wavenumber. Samples were scanned from 450 to 4000 cm^{-1} at a spectral resolution of 4 cm^{-1}. The morphology of the MgO particulate samples was examined with a scanning electron microscope (SEM-VEGA 3; TESCAN, Brun, Czech Republic); for the analysis, each sample was dispersed in water, and a drop of dispersion was deposited on an Al sample holder. The samples were gold-coated to render their surfaces conductive.

4.6. Antifungal and Antibacterial Tests

4.6.1. Disk Agar Diffusion Method

The MgO samples were tested against *Candida Albicans* (ATCC 64548), *Escherichia coli* (ATCC 8739), and *Staphylococcus aureus* (ATCC 6538P) by disk diffusion using a methodology adapted from the Clinical and Laboratory Standard Institute [48]. The MgO sample disks were prepared with 250 mg of each solid oxide and deposited in a steel mold (Sigma-Aldrich Z506699-1EA) and subjected to pressure of approximately 3.0 kg.cm^{-2} in a hydraulic press, forming disks of 13 mm diameter. The pathogens were cultured overnight and then diluted with saline solution (0.85%) to a concentration of 108 CFU.mL^{-1} (McFarland 0.5). The pathogen suspension was inoculated on the surface of Muller–Hinton

Agar using a sterile swab. The sample disks of MgO-St, MgO-CP, and MgO-Av were placed on the agar surface and then incubated at 36 °C (±1) for 24 h. The tests were performed in triplicate, and the antibacterial and antifungal activity was evaluated by measuring the diameters of halos of growth inhibition strains assayed (in mm). The obtained data were analyzed by one-way analysis of variance (ANOVA) and t-test analysis. A value of $p < 0.05$ was considered to be statistically significant.

4.6.2. Minimum Inhibitory Concentration (MIC) Assay

The antibacterial properties of the obtained oxides were also investigated by the minimum inhibitory concentration (MIC) test against *Staphylococcus aureus* (ATCC. 25923), *Escherichia coli* (ATCC 25922), and *Candida Albicans* (ATCC 64548). *E. coli*, *S. aureus*, and *C. albicans* inoculates were grown at 35 °C for 18 h and diluted to obtain a final well density of 10^5 CFU.mL^{-1}. Solutions of different concentrations of each oxide were prepared individually and diluted in dimethyl sulfoxide (DMSO) to reach final concentrations ranging from 400 µg.mL^{-1} to 850 µg.mL^{-1}, with a range of 50 µg.mL^{-1}. In each well, 150 µL of Mueller–Hinton broth containing the inoculum and 50 µL of each dilution of MgO-St, MgO-CP, and MgO-Av oxides were added. The broth microdilution method was used according to the methodology adapted from the Clinical and Laboratory Standards Institute Manual [21] in 96-well microplates. The microplates were incubated at 35 °C for 24 h. Bacterial growth was confirmed by adding 10 µL of sterile aqueous solution (20 mg.mL^{-1}) of triphenyl tetrazolium chloride (TTC, Inlab, Brazil) after incubation at 35 °C for 30 min. The bacteria reduced the TTC dye from yellow to red, indicating bacterial growth [49].

5. Conclusions

MgO samples were successfully obtained using three synthetic routes. The routes are efficient, reproducible, and low-cost methods that use suitable fuels from renewable sources. The single-phase periclase was identified for all MgO samples. The MgO-St sample showed crystallinity and the highest percentage of magnesium (61.7%). These results reflect the efficiency of controlling the microorganisms *Staphylococcus aureus* and *Candida albicans* seen in the agar disk diffusion method. Moreover, the MIC method showed the concentration of inhibition of the studied microorganisms. The most satisfactory results were observed for the antibacterial and antifungal tests with the MgO-St sample. This study presents three promising and low-cost synthesis methodologies for the preparation of antibacterial and antifungal MgO materials.

Supplementary Materials: The following supporting information can be downloaded at: https://www.mdpi.com/article/10.3390/molecules28010142/s1, Figure S1: Elemental analysis of the MgO samples MgO-ST, MgO-Av, and MgO-CP. The Identification was performed using the Mg kα line located at 1.25 keV; Figure S2: Elemental analysis of the organic precursors and magnesium salt. The percentage of Mg presented by the precursors was 6.6% for starch, 0.1% for *Aloe vera*, and 1.7% for citrus pectin. The Mg(NO$_3$)$_2$.6H$_2$O salt used in the three synthetic routes showed purity of 98% of magnesium. Identification was performed using the Mg kα line located at 1.25 keV.

Author Contributions: Conceptualization, N.B., F.J.A. and J.d.O.P.; methodology, N.B. and J.d.O.P.; validation, N.B., D.F.L.H. and J.d.S.C.; formal analysis, N.B., J.d.S.C., J.d.O.P., S.J. and C.B.; investigation, N.B., J.d.S.C. and C.M.Z.; resources, C.B. and F.J.A.; data curation, N.B., J.d.O.P., D.F.L.H. and J.d.S.C.; writing—original draft preparation, N.B. and J.d.O.P.; writing—review and editing, C.B., F.J.A. and C.M.Z.; visualization, N.B.; supervision, F.J.A. and C.B.; project administration, F.J.A.; funding acquisition, F.J.A. and C.B. All authors have read and agreed to the published version of the manuscript.

Funding: N.B. acknowledges CAPES (grant number 88887.628497/2021-00) for a graduate scholarship. F.J.A. is grateful for a CNPq Productivity grant (308625/2019-6) and the grant CNPq (427127/2018-1). C.B. is a research associate for FR-FNRS, Belgium.

Institutional Review Board Statement: Not applicable.

Informed Consent Statement: Not applicable.

Data Availability Statement: Not applicable.

Acknowledgments: The authors are grateful to the funding agencies: CNPq, Capes, Finep, Fundação Araucária, and FR-FNRS.

Conflicts of Interest: The authors declare no conflict of interest.

Sample Availability: Samples of the compounds magnesium oxides (MgO-St, MgO-CP and MgO-Av) are available from the authors.

References

1. Xu, C.; Yu, Z.; Yuan, K.; Jin, X.; Shi, S.; Wang, X.; Zhu, L.; Zhang, G.; Xu, D.; Jiang, H. Improved Preparation of Electrospun MgO Ceramic Fibers with Mesoporous Structure and the Adsorption Properties for Lead and Cadmium. *Ceram. Int.* **2019**, *45*, 3743–3753. [CrossRef]
2. Dhal, J.P.; Sethi, M.; Mishra, B.G.; Hota, G. MgO Nanomaterials with Different Morphologies and Their Sorption Capacity for Removal of Toxic Dyes. *Mater. Lett.* **2015**, *141*, 267–271. [CrossRef]
3. Bassioni, G.; Farid, R.; Mohamed, M.; Hammouda, R.M.; Kühn, F.E. Effect of Different Parameters on Caustic Magnesia Hydration and Magnesium Hydroxide Rheology: A Review. *Mater. Adv.* **2021**, *2*, 6519–6531. [CrossRef]
4. Karthik, K.; Dhanuskodi, S.; Prabu Kumar, S.; Gobinath, C.; Sivaramakrishnan, S. Microwave Assisted Green Synthesis of MgO Nanorods and Their Antibacterial and Anti-Breast Cancer Activities. *Mater. Lett.* **2017**, *206*, 217–220. [CrossRef]
5. Karthik, K.; Dhanuskodi, S.; Gobinath, C.; Prabukumar, S.; Sivaramakrishnan, S. Fabrication of MgO Nanostructures and Its Efficient Photocatalytic, Antibacterial and Anticancer Performance. *J. Photochem. Photobiol. B Biol.* **2019**, *190*, 8–20. [CrossRef] [PubMed]
6. Seif, S.; Marofi, S.; Mahdavi, S. Removal of Cr3+ Ion from Aqueous Solutions Using MgO and Montmorillonite Nanoparticles. *Environ. Earth Sci.* **2019**. [CrossRef]
7. Ahmed, S.; Guo, Y.; Huang, R.; Li, D.; Tang, P.; Feng, Y. Hexamethylene Tetramine-Assisted Hydrothermal Synthesis of Porous Magnesium Oxide for High-Efficiency Removal of Phosphate in Aqueous Solution. *J. Environ. Chem. Eng.* **2017**, *5*, 4649–4655. [CrossRef]
8. Makhluf, S.; Dror, R.; Nitzan, Y.; Abramovich, Y.; Jelinek, R.; Gedanken, A. Microwave-Assisted Synthesis of Nanocrystalline MgO and Its Use as a Bacteriocide. *Adv. Funct. Mater.* **2005**, *15*, 1708–1715. [CrossRef]
9. Wu, S.; Dong, Y.; Li, X.; Gong, M.; Zhao, R.; Gao, W.; Wu, H.; He, A.; Li, J.; Wang, X.; et al. Microstructure and Magnetic Properties of FeSiCr Soft Magnetic Powder Cores with a MgO Insulating Layer Prepared by the Sol-Gel Method. *Ceram. Int.* **2022**, *48*, 22237–22245. [CrossRef]
10. Primo, J.d.O.; Bittencourt, C.; Acosta, S.; Sierra-Castillo, A.; Colomer, J.F.; Jaerger, S.; Teixeira, V.C.; Anaissi, F.J. Synthesis of Zinc Oxide Nanoparticles by Ecofriendly Routes: Adsorbent for Copper Removal From Wastewater. *Front. Chem.* **2020**, *8*, 571790. [CrossRef]
11. Brito, G.F.; Agrawal, P.; Araujo, E.M.; Melo, J.J.A. Biopolímeros, Polímeros Biodegradáveis e Polímeros Verdes. *Revista Eletrônica de Materiais e Processos. Rev. Eletrônica Mater. e Process.* **2011**, *6*, 127–139.
12. Dalpasquale, M.; Mariani, F.Q.; Müller, M.; Anaissi, F.J. Citrus Pectin as a Template for Synthesis of Colorful Aluminates. *Dye. Pigment.* **2016**, *125*, 124–131. [CrossRef]
13. Choi, S.; Chung, M.H. A Review on the Relationship between Aloe Vera Components and Their Biologic Effects. *Semin. Integr. Med.* **2003**, *1*, 53–62. [CrossRef]
14. Hamman, J.H. Composition and Applications of Aloe Vera Leaf Gel. *Molecules* **2008**, *13*, 1599–1616. [CrossRef] [PubMed]
15. Ferreira, P.P.L.; Melo, D.M.d.A.; Medeiros, R.L.B.d.A.; de Araújo, T.R.; Maziviero, F.V.; de Oliveira, Â.A.S. Green Synthesis with Aloe Vera of MgAl2O4 Substituted by Mn and without Calcination Treatment. *Res. Soc. Dev.* **2022**, *11*, e14411628873. [CrossRef]
16. Strathdee, S.A.; Davies, S.C.; Marcelin, J.R. Confronting Antimicrobial Resistance beyond the COVID-19 Pandemic and the 2020 US Election. *Lancet* **2020**, *396*, 1050–1053. [CrossRef] [PubMed]
17. Williams, D.W.; Jordan, R.P.C.; Wei, X.-Q.; Alves, C.T.; Wise, M.P.; Wilson, M.J.; Lewis, M.A.O. Interactions of Candida Albicans with Host Epithelial Surfaces. *J. Oral Microbiol.* **2013**, *5*, 22434. [CrossRef]
18. Aliprandini, E.; Tavares, J.; Panatieri, R.H.; Thiberge, S.; Yamamoto, M.M.; Silvie, O.; Ishino, T.; Yuda, M.; Dartevelle, S.; Traincard, F.; et al. Cytotoxic Anti-Circumsporozoite Antibodies Target Malaria Sporozoites in the Host Skin. *Nat. Microbiol.* **2018**, *3*, 1224–1233. [CrossRef]
19. Hayat, S.; Muzammil, S.; Rasool, M.H.; Nisar, Z.; Hussain, S.Z.; Sabri, A.N.; Jamil, S. In Vitro Antibiofilm and Anti-Adhesion Effects of Magnesium Oxide Nanoparticles against Antibiotic Resistant Bacteria. *Microbiol. Immunol.* **2018**, *62*, 211–220. [CrossRef]
20. Bindhu, M.R.; Umadevi, M.; Kavin Micheal, M.; Arasu, M.V.; Abdullah Al-Dhabi, N. Structural, Morphological and Optical Properties of MgO Nanoparticles for Antibacterial Applications. *Mater. Lett.* **2016**, *166*, 19–22. [CrossRef]
21. Lowy, F.D. Antimicrobial Resistance: The Example of Staphylococcus Aureus. *J. Clin. Invest.* **2003**, *111*, 1265–1273. [CrossRef] [PubMed]

22. Cowen, L.E.; Anderson, J.B.; Kohn, L.M. Evolution of Drug Resistance in Candida Albicans. *Annu. Rev. Microbiol.* **2002**, *56*, 139–165. [CrossRef] [PubMed]
23. Masterson, K.; Meade, E.; Garvey, M.; Lynch, M.; Major, I.; Rowan, N.J. Development of a Low-Temperature Extrusion Process for Production of GRAS Bioactive-Polymer Loaded Compounds for Targeting Antimicrobial-Resistant (AMR) Bacteria. *Sci. Total Environ.* **2021**, *800*, 149545. [CrossRef] [PubMed]
24. Shakir Shnain Al-Shammari, R.; Kareem Hammood Jaberi, A.; Lungu, A.; Brebenel, I.; Khaled Kaddem Al-Sudani, W.; Hafedh Mohammed Al-Saedi, J.; Pop, C.E.; Mernea, M.; Stoian, G.; Mihailescu, D.F. The Antifungal Activity of Zein Nanoparticles Loaded With Transition Metal Ions. *Rev. Roum. Chim.* **2021**, *66*, 829–834. [CrossRef]
25. Umaralikhan, L.; Jamal Mohamed Jaffar, M. Green Synthesis of MgO Nanoparticles and It Antibacterial Activity. *Iran. J. Sci. Technol. Trans. A Sci.* **2018**, *42*, 477–485. [CrossRef]
26. Mutis González, N.; Pineda Gómez, P.; Rodríguez García, M.E. Effect of the Addition of Potassium and Magnesium Ions on the Thermal, Pasting, and Functional Properties of Plantain Starch (Musa Paradisiaca). *Int. J. Biol. Macromol.* **2019**, *124*, 41–49. [CrossRef]
27. Mousavi, S.M.; Hashemi, S.A.; Ramakrishna, S.; Esmaeili, H.; Bahrani, S.; Koosha, M.; Babapoor, A. Green Synthesis of Supermagnetic Fe3O4–MgO Nanoparticles via Nutmeg Essential Oil toward Superior Anti-Bacterial and Anti-Fungal Performance. *J. Drug Deliv. Sci. Technol.* **2019**, *54*, 101352. [CrossRef]
28. Moulavi, M.H.; Kale, B.B.; Bankar, D.; Amalnerkar, D.P.; Vinu, A.; Kanade, K.G. Green Synthetic Methodology: An Evaluative Study for Impact of Surface Basicity of MnO2 Doped MgO Nanocomposites in Wittig Reaction. *J. Solid State Chem.* **2019**, *269*, 167–174. [CrossRef]
29. Li, N.; Dang, H.; Chang, Z.; Zhao, X.; Zhang, M.; Li, W.; Zhou, H.; Sun, C. Synthesis of Uniformly Distributed Magnesium Oxide Micro-/Nanostructured Materials with Deep Eutectic Solvent for Dye Adsorption. *J. Alloys Compd.* **2019**, *808*. [CrossRef]
30. Siddqui, N.; Sarkar, B.; Pendem, C.; Khatun, R.; Sivakumar Konthala, L.N.; Sasaki, T.; Bordoloi, A.; Bal, R. Highly Selective Transfer Hydrogenation of α,β-Unsaturated Carbonyl Compounds Using Cu-Based Nanocatalysts. *Catal. Sci. Technol.* **2017**, *7*, 2828–2837. [CrossRef]
31. Zahir, M.H.; Rahman, M.M.; Irshad, K.; Rahman, M.M. Shape-Stabilized Phase Change Materials for Solar Energy Storage: MgO and Mg(OH)2 Mixed with Polyethylene Glycol. *Nanomaterials* **2019**, *9*. [CrossRef] [PubMed]
32. Nga, N.K.; Thuy Chau, N.T.; Viet, P.H. Preparation and Characterization of a Chitosan/MgO Composite for the Effective Removal of Reactive Blue 19 Dye from Aqueous Solution. *J. Sci. Adv. Mater. Devices* **2020**, *5*, 65–72. [CrossRef]
33. Kumar, A.; Kumar, J. Defect and Adsorbate Induced Infrared Modes in Sol-Gel Derived Magnesium Oxide Nano-Crystallites. *Solid State Commun.* **2008**, *147*, 405–408. [CrossRef]
34. Mageshwari, K.; Mali, S.S.; Sathyamoorthy, R.; Patil, P.S. Template-Free Synthesis of MgO Nanoparticles for Effective Photocatalytic Applications. *Powder Technol.* **2013**, *249*, 456–462. [CrossRef]
35. Ananda, A.; Ramakrishnappa, T.; Archana, S.; Yadav, L.S.R.; Shilpa, B.M.; Nagaraju, G.; Jayanna, B.K. Materials Today: Proceedings Green Synthesis of MgO Nanoparticles Using Phyllanthus Emblica for Evans Blue Degradation and Antibacterial Activity. *Mater Today Proc.* **2021**, *49*, 801–810. [CrossRef]
36. Mastuli, M.S.; Kamarulzaman, N.; Nawawi, M.A.; Mahat, A.M.; Rusdi, R.; Kamarudin, N. Growth Mechanisms of MgO Nanocrystals via a Sol-Gel Synthesis Using Different Complexing Agents. *Nanoscale Res. Lett.* **2014**, *9*, 134. [CrossRef]
37. Guo, L.; Lei, R.; Zhang, T.C.; Du, D.; Zhan, W. Insight into the Role and Mechanism of Polysaccharide in Polymorphous Magnesium Oxide Nanoparticle Synthesis for Arsenate Removal. *Chemosphere* **2022**, *296*, 133878. [CrossRef]
38. Sejas, L.M.; Silbert, S.; Reis, A.O.; Sader, H.S. Avaliação Da Qualidade Dos Discos Com Antimicrobianos Para Testes de Disco-Difusão Disponíveis Comercialmente No Brasil. *J. Bras. Patol. e Med. Lab.* **2003**, *39*. [CrossRef]
39. Sathishkumar, S.; Sridevi, C.; Rajavel, R.; Karthikeyan, P. Journal of Science: Advanced Materials and Devices Smart Fl Ower like MgO / Tb, Eu-Substituted Hydroxyapatite Dual Layer Coating on 316L SS for Enhanced Corrosion Resistance, Antibacterial Activity and Osteocompatibility. *J. Sci. Adv. Mater. Devices* **2020**, *10*. [CrossRef]
40. Medina-Ramírez, I.E.; Arzate-Cardenas, M.A.; Mojarro-Olmos, A.; Romo-López, M.A. Synthesis, Characterization, Toxicological and Antibacterial Activity Evaluation of Cu@ZnO Nanocomposites. *Ceram. Int.* **2019**, *45*, 17476–17488. [CrossRef]
41. EUCAST Definitive Document. Methods for the Determination of Susceptibility of Bacteria to Antimicrobial Agents.Terminol. *Clin. Microbiol. Infect* **1998**, *4*, 291–296. [CrossRef]
42. Almeida, E.; De Bona, M.; Gisele, F.; Fruet, T.K.; Cristina, T.; Jorge, M.; De Moura, A.C. Comparação de Métodos Para Avaliação Da Atividade Antimicrobiana e Determinação Da Concentração Inibitória Mínima (Cim) de Extratos Vegetais Aquosos e Etanólicos. *Arq. Inst. Biol.* **2014**, *81*, 218–225. [CrossRef]
43. Moreno, S.; Scheyer, T.; Romano, C.S. Antioxidant and Antimicrobial Activities of Rosemary Extracts Linked to Their Polyphenol Composition Antioxidant and Antimicrobial Activities of Rosemary Extracts Linked to Their Polyphenol Composition. *Free Radic. Res.* **2016**, *40*, 223–231. [CrossRef] [PubMed]
44. El-Shaer, A.; Abdelfatah, M.; Mahmoud, K.R.; Momay, S.; Eraky, M.R. Correlation between Photoluminescence and Positron Annihilation Lifetime Spectroscopy to Characterize Defects in Calcined MgO Nanoparticles as a First Step to Explain Antibacterial Activity. *J. Alloys Compd.* **2020**, *817*, 152799. [CrossRef]
45. Almontasser, A.; Parveen, A.; Azam, A. Synthesis, Characterization and Antibacterial Activity of Magnesium Oxide (MgO) Nanoparticles. *IOP Conf. Ser. Mater. Sci. Eng.* **2019**, *577*. [CrossRef]

46. Primo, J.d.O.; Borth, K.W.; Peron, D.C.; Teixeira, V.d.C.; Galante, D.; Bittencourt, C.; Anaissi, F.J. Synthesis of Green Cool Pigments (CoxZn1-XO) for Application in NIR Radiation Reflectance. *J. Alloys Compd.* **2019**, *780*, 17–24. [CrossRef]
47. Khort, A.; Hedberg, J.; Mei, N.; Romanovski, V.; Blomberg, E.; Odnevall, I. Corrosion and Transformation of Solution Combustion Synthesized Co, Ni and CoNi Nanoparticles in Synthetic Freshwater with and without Natural Organic Matter. *Sci. Rep.* **2021**, *11*, 7860. [CrossRef]
48. *M07*; Methods for Dilution Antimicrobial Susceptibility Tests for Bacteria That Grow Aerobically. Clinical and Laboratory Standards Institute: Wayne, PA, USA, 2006; Volume 32.
49. Radaelli, M.; da Silva, B.P.; Weidlich, L.; Hoehne, L.; Flach, A.; da Costa, L.A.M.A.; Ethur, E.M. Antimicrobial Activities of Six Essential Oils Commonly Used as Condiments in Brazil against Clostridium Perfringens. *Braz. J. Microbiol.* **2016**, *47*, 424–430. [CrossRef]

Disclaimer/Publisher's Note: The statements, opinions and data contained in all publications are solely those of the individual author(s) and contributor(s) and not of MDPI and/or the editor(s). MDPI and/or the editor(s) disclaim responsibility for any injury to people or property resulting from any ideas, methods, instructions or products referred to in the content.

Article

Visible-Light-Enhanced Antibacterial Activity of Silver and Copper Co-Doped Titania Formed on Titanium via Chemical and Thermal Treatments

Kanae Suzuki [1], Misato Iwatsu [2], Takayuki Mokudai [3], Maiko Furuya [2], Kotone Yokota [2], Hiroyasu Kanetaka [2], Masaya Shimabukuro [4], Taishi Yokoi [4] and Masakazu Kawashita [4,*]

[1] Graduate School of Biomedical Engineering, Tohoku University, 6-6-12 Aramaki-Aoba, Aoba-ku, Sendai 980-8579, Japan
[2] Graduate School of Dentistry, Tohoku University, 4-1, Seiryo-machi, Aoba-ku, Sendai 980-8575, Japan
[3] Institute of Materials Research, Tohoku University, 2-1-1, Katahira, Aoba-ku, Sendai 980-8577, Japan
[4] Institute of Biomaterials and Bioengineering, Tokyo Medical and Dental University, 2-3-10 Kanda-surugadai, Chiyoda-ku, Tokyo 101-0062, Japan
* Correspondence: kawashita.bcr@tmd.ac.jp; Tel.: +81-3-5280-8016

Abstract: Dental implants made of titanium (Ti) are used in dentistry, but peri-implantitis is a serious associated problem. Antibacterial and osteoconductive Ti dental implants may decrease the risk of peri-implantitis. In this study, titania (TiO_2) co-doped with silver (Ag) at 2.5 at.% and copper (Cu) at 4.9 at.% was formed on Ti substrates via chemical and thermal treatments. The Ag and Cu co-doped TiO_2 formed apatite in a simulated body fluid, which suggests osteoconductivity. It also showed antibacterial activity against *Escherichia coli*, which was enhanced by visible-light irradiation. This enhancement might be caused by the synergistic effect of the release of Ag and Cu and the generation of •OH from the sample. Dental implants with such a Ag and Cu co-doped TiO_2 formed on their surface may reduce the risk of peri-implantitis.

Keywords: titania; silver; copper; antibacterial activity; visible-light-responsive photocatalysis

1. Introduction

Dental implants made of titanium (Ti) are widely used in dentistry, but peri-implantitis [1–4], which has a prevalence rate of about 22% [5], is a serious problem. The incidence of peri-implantitis caused by Ti dental implants can be decreased by inducing antibacterial activity via the control of surface topology [6], the incorporation of antibacterial metals [7], or a functional layer coating [8]. One strategy for preparing antibacterial Ti dental implants is the formation of a titanium oxide (TiO_2) layer with photocatalytic antibacterial activity [9–11] on their surfaces. For example, Suketa et al. reported the photocatalytic antibacterial activity of TiO_2 film formed on Ti via plasma source ion implantation [12]. It has been reported that a TiO_2 layer formed on Ti via chemical and thermal treatments can form apatite on its surface in a simulated body fluid (SBF) [13] and bond to living bone [14]. Therefore, Ti dental implants with TiO_2 formed on their surfaces are expected to exhibit photocatalytic antibacterial activity as well as bone-bonding ability. However, for such Ti dental implants, the photocatalytic antibacterial activity of TiO_2 is exhibited only under exposure to short-wavelength invisible light such as ultraviolet light, which is toxic to living organisms.

When TiO_2 is doped with elements such as nitrogen (N) [15,16] and copper (Cu) [17,18], it can show photocatalytic activity even under visible light. Several mechanisms for the visible-light-responsive photocatalytic activity of Cu-doped TiO_2 have been proposed, depending on the chemical state of Cu [19], such as the surface plasmon resonance effect of Cu nanoparticles [20] and electron transfer from TiO_2 to CuO [21,22] or from Cu_2O to TiO_2 [23,24]. It has been reported that 650 nm light from a light-emitting diode (LED) penetrates the gingiva and activates the photosensitizer within the gingival sulcus to kill bacteria that reside around the gingival sulcus [25]. Therefore, Ti dental implants with doped TiO_2 on their surfaces can reduce the risk of peri-implantitis with periodic or on-demand irradiation of visible light at a dental clinic. We previously prepared N-doped TiO_2 [26–29] and Cu-doped TiO_2 [30] on Ti and investigated their surface structure, apatite formation ability in an SBF, and antibacterial activity. However, it is necessary to improve the antibacterial activity of N-doped or Cu-doped TiO_2. One possible approach to achieve this is to increase the N or Cu content, but our method limits the amount of N or Cu that can be doped into TiO_2 to improve photocatalytic antibacterial activity and apatite formation ability [28,31].

Therefore, in this study, we tried to co-dope silver (Ag) and Cu into TiO_2. The excellent antibacterial properties of Ag are expected to improve the antibacterial activity of the dental implants with or without visible-light irradiation. The antibacterial activity of samples is discussed in terms of their photocatalytic activity and the release of Ag and Cu from the samples. The present findings will contribute to the development of dental implants with antibacterial activity to prevent peri-implantitis with and without visible-light irradiation.

2. Results and Discussion

A network-like structure formed on the surfaces of both AG-CU and AG, whereas small particles formed only on the surface of AG (Figure 1a). A similar network-like structure with small particles was previously reported [32–34]. The network-like structure was composed of anatase, rutile, and metallic silver (Figure 1b). The intensity of the TF-XRD peak attributed to metallic silver around the 2θ angle of 44° was much higher for AG than for AG-CU, which suggests that the small particles on the surface of AG were mainly composed of metallic silver. The intensity of the TF-XRD peak attributed to rutile at the 2θ angle of around 27° was larger than that attributed to anatase at the 2θ angle of around 25° for AG-CU; the opposite result was obtained for AG. This indicates that rutile and anatase preferentially formed on AG-CU and AG, respectively. The preferential formation of rutile on AG-CU was likely caused by Cu, a dopant that promotes the phase transformation of anatase to rutile [35].

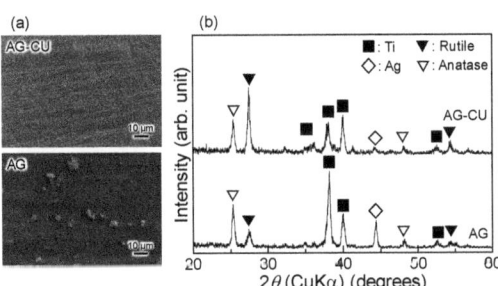

Figure 1. (a) SEM images and (b) TF-XRD patterns of samples.

AG-CU contained Ag at 2.5 at.% and Cu at 4.9 at.% on its surface, and AG contained Ag at 6.3 at.% on its surface (Table 1). AG-CU contained almost twice as much Cu as Ag. The amount of Ag in AG was higher than that in AG-CU, which can be attributed to the higher concentration of Ag in the silver nitrate ($AgNO_3$) solution used for the treatment of AG ($\cong$1 mM) compared to that ($\cong$0.5 mM) used for the treatment of AG-CU. Although the concentrations of Ag and Cu in the $AgNO_3$-$Cu(NO_3)_2$ mixed solution used for the treatment of AG-CU were the same ($\cong$0.5 mM), AG-CU contained almost twice as much Cu as Ag on its surface. These results indicate that Ag and Cu can be co-doped into a sample by using an $AgNO_3$-$Cu(NO_3)_2$ mixed solution, but the amount of Ag doped into the sample will not be simply proportional to the Ag concentration of the Ag- and Cu-containing solution used for treatment.

Table 1. Surface composition of samples (mean ± SD).

Sample	Composition (at.%)				
	O	Ti	Ag	Cu	C
AG-CU	59.2 ± 0.6	31.0 ± 0.8	2.5 ± 0.2	4.9 ± 0.6	2.3 ± 0.2
AG	59.0 ± 1.2	31.7 ± 0.7	6.3 ± 1.8	—	3.1 ± 0.3

—: not measured.

Figure 2 shows the Ag 3d and Cu 2p electron energy region spectra of the samples and Table 2 summarizes the binding energy (E_B) and modified Auger parameter (α') values of the samples. The chemical states of Ag and Cu can be determined from a comparison of E_B with α' on the Wagner plot. The Ag $3d_{5/2}$ peak around 368.5 eV and the α' value of around 723.4 eV for AG-CU and AG suggest that Ag mainly existed in an oxide state on their surfaces [36]. Taking into account that the TF-XRD peaks of metallic silver were observed for AG-CU and AG (Figure 1b), we speculate that the surface of the metallic silver was oxidized in AG-CU and AG.

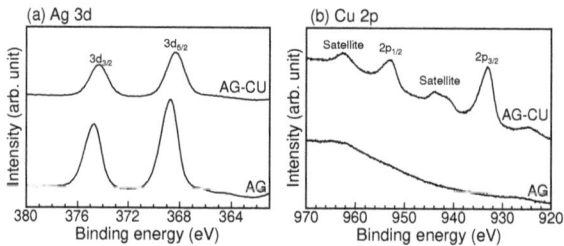

Figure 2. (a) Ag 3d and (b) Cu 2p electron energy region spectra of samples.

Table 2. Summary of binding energy (E_B) and modified Auger parameter (α') values of samples (mean ± SD).

Sample	Element	E_B (eV)	α' (eV)
AG-CU	Ag	368.4 ± 0.2	723.5 ± 0.4
	Cu	933.0 ± 0.2	1850.4 ± 0.2
AG	Ag	368.6 ± 0.3	723.3 ± 0.1

Cu $2p_{3/2}$ peaks were observed at around 933.0 eV for AG-CU, whereas no Cu $2p_{3/2}$ peak was observed for AG. The Cu $2p_{3/2}$ peak around 933.0 eV and the α' value around 1850.4 eV suggest that Cu mainly existed as Cu_2O on the surface of AG-CU [37]. These results indicate that copper was successfully doped into the sample surface by the present surface treatments. The lack of a TF-XRD peak corresponding to copper compounds for AG-CU (Figure 1b) and the apparent Cu 2p peak (Figure 2b and Table 1) indicate that the crystallinity of the doped copper was low for both samples. The formation of Cu_2O with low crystallinity in AG-CU is interesting, but its mechanism is unclear. This topic is worthy of further investigation.

The SEM images of the samples after immersion in the SBF (Figure 3) indicate that apatite uniformly formed on the surface of AG, whereas it partially formed on the surface of AG-CU. This difference in apatite formation ability between these samples is consistent with the intensity of the TF-XRD peak of apatite at the 2θ angle of around 32° being much smaller for AG-CU than for AG. The relationship between apatite formation ability in an SBF and the surface structure of TiO_2 formed on Ti [13,31,38–41] or TiO_2 gels [42] is not fully understood; nevertheless, in this study, the higher formation of anatase compared to that of rutile in AG (Figure 1b) may be responsible for the better apatite formation ability of AG.

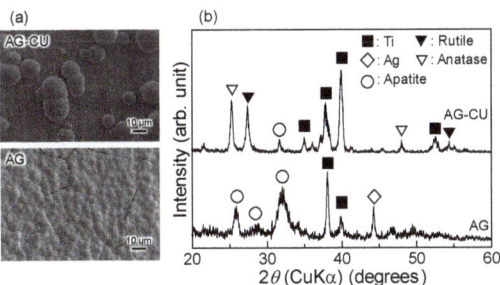

Figure 3. (**a**) SEM images and (**b**) TF-XRD patterns of samples after immersion in SBF for 7 days.

A slightly higher amount of Ag was released from AG-CU than from AG. The Ag concentration in PBS reached around 8 µM for 3 days (Figure 4a). However, the Ag concentration was saturated at around 7 days, which indicates that the release of Ag from AG-CU almost stopped at around 7 days. In contrast, AG released Ag gradually and continuously for 28 days. AG-CU slowly released Cu; the Cu concentration reached around 3 µM by day 28. These results indicate that AG-CU preferentially releases Ag over Cu, but the release of Ag is almost stopped at around 7 days even though AG continuously releases Ag for 28 days. Figure 4b was obtained by plotting the Ag and Cu concentrations against the square root of the soaking period. The concentration of Ag released from AG-CU within 3 days and that released from AG within 28 days are proportional to the square root of the soaking period. This result suggests that Ag was released from both samples via ion exchange [30,43], although the rate and duration of Ag release were different between the samples. The concentration of Cu released from AG-CU within 28 days is also proportional to the square root of the soaking period, which suggests that Cu was released from AG-CU via ion exchange [44]. However, the mechanism of Ag and Cu release from samples should be further investigated because Ag and Cu were mainly present as metallic silver with an oxidized surface and Cu_2O, respectively (Figures 1 and 2, and Table 2), and they are not likely to be released via ion exchange. The slightly more rapid release of Ag from AG-CU than from AG and the continuous release of Cu from AG-CU may lead to antibacterial activity that is somewhat strong at the initial stage of implantation and continues for a long period.

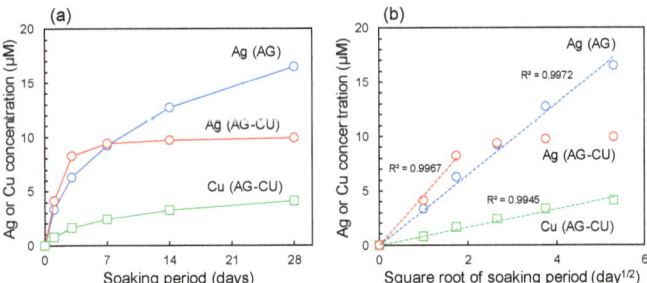

Figure 4. Ag and Cu ion release behavior from samples in PBS. (**a**) Accumulated-released amounts of Ag or Cu vs. soaking period and (**b**) accumulated-released amounts of Ag or Cu vs. square root of soaking period.

Without visible-light irradiation, the number of viable bacteria was significantly smaller for AG and AG-CU than for untreated Ti, and slightly smaller for AG-CU than for AG (Figure 5). The rapid release of Ag and sustained release of Cu from AG-CU (Figure 4) might be responsible for the higher antibacterial activity of AG-CU compared to that of AG. The number of viable bacteria was significantly decreased by visible-light irradiation for

AG-CU and AG compared to untreated Ti, and AG-CU showed extremely strong antibacterial activity under visible-light irradiation. The number of viable bacteria on untreated Ti (control) decreased under visible-light irradiation. Although an LED generates much less heat than a conventional incandescent bulb, the decrease in the number of viable bacteria on untreated Ti under visible-light irradiation may be attributed to the heat generated by the LED light, which was placed only 10 cm from the sample and had a high intensity of 250 W·m^{-2}. Here, we briefly discuss the changes in the oxidation state of copper after antibacterial activity testing. Although XPS spectra of AG-CU after antibacterial testing should be measured to clarify the change in oxidation state of copper in the future, it is possible that Cu^{2+} is formed from Cu_2O on the surface of AG-CU after antibacterial testing because the proportion of Cu^{2+} on the surface of copper metal increases after soaking in a bacteria-containing solution, and Cu^{2+} is the most stable chemical state against corrosion and bacteria [37].

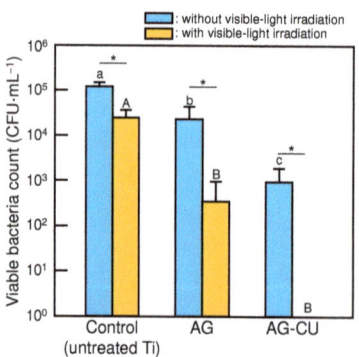

Figure 5. Number of viable bacteria for samples under conditions with and without visible-light irradiation. Bars with different letters (lowercase a–c for no visible-light irradiation group and uppercase A and B for visible-light irradiation group) are significantly different ($p < 0.01$). Asterisk (*) represents significant differences ($p < 0.01$) between no visible-light irradiation and visible-light irradiation.

Next, the antibacterial activity of AG-CU under visible-light irradiation is discussed in terms of the generation of ROS. The concentration of the hydroxyl radical (•OH) was measured via ESR using DMPO as the spin-trapping agent. Peaks of DMPO-OH were observed for AG-CU and AG. The intensity of the peaks was larger for AG-CU than for AG (Figure S1). Table 3 shows the concentrations of H_2O_2 and •OH for the samples. The H_2O_2 concentrations for all samples were less than 0.1 µM, much lower than the H_2O_2 concentrations (>1.25 µM) that can effectively kill *E. coli* [45–47]. The •OH concentration was higher for AG-CU than for AG and the control. Therefore, •OH radicals are likely to be generated by a reaction between hydroxide ions (OH^-) and holes (h^+), $OH^- + h^+ \rightarrow$ •OH, namely a direct photocatalytic effect, on the surface of AG-CU. The generated •OH may contribute to the antibacterial activity of AG-CU under visible-light irradiation. In summary, it is thought that the excellent antibacterial activity of AG-CU under visible-light irradiation (Figure 5) can be attributed to the synergistic effect of the release of Ag and Cu (Figure 4) and the generation of •OH from the sample (Table 3). The details of the synergistic effect are still unclear, but it is possible that bacteria damaged by released Ag and Cu are more likely to be killed by •OH, or vice versa.

Table 3. Concentrations of hydrogen peroxide (H_2O_2) and hydroxyl radical (•OH) for samples.

Sample	Concentration (µM)	
	H_2O_2	•OH
AG-CU	8.0×10^{-2}	2.5
AG	8.9×10^{-2}	1.3
Control (untreated Ti)	8.9×10^{-2}	1.4

3. Materials and Methods

3.1. Sample Preparation

A commercially pure Ti chip with dimensions of 10 mm × 10 mm × 1 mm (purity: 99.9%, TIE04CB, Kojundo Chemical Lab. Co., Ltd., Saitama, Japan) was used as the original substrate and polished using a diamond pad (no. 400, Maruto Instrument Co., Ltd., Tokyo, Japan). The polished Ti chip was ultrasonically washed once with acetone (99%, Nacalai Tesque, Inc., Kyoto, Japan) and twice with ultrapure water for 10 min. The washed chip was dried at room temperature and atmospheric pressure. Subsequently, an aqueous NaOH solution was prepared by dissolving 1.031 g of NaOH (FUJIFILM Wako Pure Chemical Corp., Osaka, Japan) in 5 mL of ultrapure water. The washed chip was immersed in the NaOH aqueous solution in a round-bottomed polytetrafluoroethylene (PTFE) test tube with a cap (code 04936, SANPLATEC Corp., Osaka, Japan). The test tube was shaken at 120 strokes·min^{-1} for 24 h at 60 °C using a shaking bath. After the completion of the NaOH treatment, the Ti chip was removed from the test tube and washed with ultrapure water to obtain the NaOH-treated Ti chip. Subsequently, 0.085 g of silver nitrate ($AgNO_3$, FUJIFILM Wako Pure Chemical Corp.) was dissolved in 5 mL of ultrapure water. The $AgNO_3$ solution was diluted 100-fold to obtain approximately 1 mol·m^{-3} of $AgNO_3$ solution. In addition, 0.121 g of $Cu(NO_3)_2 \cdot 3H_2O$ (FUJIFILM Wako Pure Chemical Corp.) was dissolved in 5 mL of ultrapure water. The $Cu(NO_3)_2$ solution was diluted 100-fold to obtain approximately 1 mol·m^{-3} of $Cu(NO_3)_2$ solution. A total of 3 mL of the diluted $AgNO_3$ solution was mixed with 3 mL of the diluted $Cu(NO_3)_2$ solution and transferred to a round-bottomed PTFE test tube with a cap. The NaOH-treated Ti chip was then immersed in this mixture and shaken at 120 strokes·min^{-1} for 48 h at 80 °C. After the treatment, the chip was removed and washed with ultrapure water. The Ti chip treated with the $AgNO_3$-$Cu(NO_3)_2$ mixed solution was heat-treated at 600 °C for 1 h using a muffle furnace (MSFS-1218, Yamada Denki Co., Ltd., Tokyo, Japan). The samples thus obtained are denoted as AG-CU. As a reference, the NaOH-treated Ti chips were immersed in 6 mL of the diluted $AgNO_3$ solution in a round-bottomed PTFE test tube with a cap, and then heat-treated at 600 °C for 1 h. The samples thus obtained are denoted as AG.

3.2. Surface Structure Analysis

The surface morphology of the samples was observed using scanning electron microscopy (SEM; VE8800, Keyence Corp., Osaka, Japan). The crystalline phase of the surface layer formed by the solution and heat treatments was characterized using thin-film X-ray diffraction (TF-XRD; RINT2200VL, Rigaku Corporation, Tokyo, Japan) with Cu Kα radiation. The composition of the surface layer was evaluated using X-ray photoelectron spectroscopy (XPS; JPS-9010MC, JEOL, Tokyo, Japan). The X-ray source was monochromatic Mg Kα radiation (1253.6 eV) at 10 kV and 10 mA. The binding energy was calibrated using the C 1s photoelectron peak at 285.0 eV as a reference. XPS peak analysis was performed using CasaXPS (version 2.3.24, Casa Software Ltd., Devon, UK). The Shirley background was subtracted from all spectra prior to fitting. The surface composition was calculated from the XPS spectra using relative sensitivity factors obtained from the CasaXPS software library (C 1s, 1.0, O 1s, 2.93; Ti $2p_{3/2}$, 5.22; Ag $3d_{5/2}$ 10.68, Cu $2p_{3/2}$ 16.73). In addition, the modified Auger parameters (α') of Ag and Cu were calculated from the Ag $3d_{5/2}$ and Ag M_4VV peaks and from the Cu $2p_{3/2}$ and Cu L_3VV peaks, respectively.

3.3. Evaluation of Apatite Formation Ability

The apatite formation ability of samples was evaluated using an SBF [48] that contained ions at concentrations (Na^+: 142.0 mM; K^+: 5.0 mM; Ca^{2+}: 2.5 mM; Mg^{2+}: 1.5 mM; Cl^-: 147.8 mM; HCO_3^-: 4.2 mM; HPO_4^{2-}: 1.0 mM; SO_4^{2-}: 0.5 mM) nearly identical to those found in human blood plasma. The SBF was prepared according to the ISO 23317:2014 protocol. All chemicals used in the preparation of the SBF were purchased from Nacalai Tesque, Inc., Kyoto, Japan. An amount of 30 mL of the prepared SBF was poured into a centrifuge tube (ECK-50ML-R, AS-ONE Corp., Osaka, Japan). The samples were immersed in the SBF and kept at 36.5 °C. After 7 days, the samples were removed from the SBF, gently washed with ultrapure water, and dried at approximately 25 °C and atmospheric pressure. The lower surface of each sample was subjected to surface analysis using SEM and TF-XRD.

3.4. Ag and Cu Ion Release Behavior

To investigate the Ag and Cu ion release behavior of each sample, 10 mL of phosphate-buffered saline (PBS, 166-23555, FUJIFILM Wako Pure Chemical Corp.) was placed in a centrifuge tube (ECK-50ML-R, AS-ONE Corp.). The sample ($n = 3$) was immersed in PBS at 36.5 °C. The PBS was refreshed at appropriate periods. The accumulated and released amounts of Ag and Cu ions from the samples at 1, 3, 7, 14, and 28 days were calculated based on the Ag and Cu concentrations in the PBS, respectively, which were measured using inductively coupled plasma atomic emission spectroscopy (ICP-AES, iCAP600, Thermo Fisher Scientific Co., Ltd., Kanagawa, Japan).

3.5. Evaluation of Antimicrobial Activity

A nutrient agar was used in petri dishes (Falcon® plastic dish for general bacteria, Corning Inc., New York, NY, USA) in 15 mL aliquots. Physiological saline was prepared by dissolving 8.5 g of sodium chloride (NaCl, Nacalai Tesque, Inc.) into 1 L of ultrapure water, which was used after sterilization at 121 °C for 20 min using a high-pressure steam sterilizer. *Escherichia coli* (*E. coli*, JCM5491) was used as the test bacterial strain. It was used after being cultured on the nutrient agar medium at 37 °C for 24 h. The bacterial mass of the cultured *E. coli* was taken with a platinum loop and dispersed in physiological saline to prepare a stock bacterial suspension ($\cong 10^8$ CFU·mL^{-1}). This stock suspension was diluted with a nutrient liquid medium to obtain a test bacterial suspension ($\cong 10^7$ CFU·mL^{-1}). The bacterial test was carried out for each sample ($n = 4$). A cell strainer (Corning Inc.) attached to a 6-well plate was used for setting the sample. The sample was placed on the cell strainer with the sample surface facing upward and 10 µL of the test bacterial suspension was dropped onto the sample. Subsequently, the sample surface was covered with a plastic film (9 mm × 9 mm × 0.06 mm) to achieve close contact. To reduce the effects of increasing temperature and drying during visible-light irradiation on the bacteria, a cooler was placed behind the 6-well plate and 1.5 mL of pure water was added to the wells to prevent the sample from drying. LED light (460 nm; SPA-10SW, Hayashi Clock Industry Co., Ltd., Tokyo, Japan) was used as the light source. The distance from the lower part of the lens to the sample surface was 10 cm, the irradiance was 250 W·m^{-2}, and the irradiation period was 30 min. This irradiation period was set under the assumption of visible-light irradiation to the abutment of dental implants during dental treatments. As a control experiment, an antibacterial test without visible-light irradiation was also conducted. A schematic illustration of the antimicrobial activity evaluation system is shown in our previous paper [30]. After either irradiation with visible light for 30 min or no irradiation for 30 min, the sample was collected together with the film, soaked in 2 mL of soybean-casein digest broth with lecithin and polysorbate 80 (SCDLP, Nihon Pharmaceutical Co. Ltd., Osaka, Japan) medium, and thoroughly stirred to wash out the bacteria. The washed-out medium was diluted 10- and 100-fold with the SCDLP medium, and 100 µL of each was seeded onto the nutrient agar medium. These media were cultured at 37 °C for 48 h. Then, the number of colonies was counted and the viable cell count was calculated. The viable bacteria count for the AG-CU and AG groups was compared by performing a one-way analysis of variance and conducting a multiple-hypothesis test (Holm's method).

3.6. Identification of Reactive Oxygen Species Induced by Visible-Light Irradiation

It is difficult to directly measure highly reactive oxygen species (ROS) and free radicals at around 25 °C. Therefore, we measured these chemical species via electron spin resonance (ESR; JES-FA-100, JEOL Ltd., Tokyo, Japan) using a spin-trapping method. 5,5-Dimethyl-1-pyrroline-N-oxide (DMPO, Labotech Co., Tokyo, Japan) was used as the spin-trapping agent. The measurement conditions were as follows: microwave power of 4.0 mW; microwave frequency of 9428.954 MHz; magnetic width of 0.1 mT; field sweep width of ±5 mT; field modulation frequency of 100 kHz; modulation width of 0.1 mT; time constant of 0.03 s; and sweep time of 0.1 min. The samples were placed in a 24-well plate and 500 µL of DMPO solution (300 mM) was added. The samples immersed in the DMPO solution were irradiated with visible light for 30 min under the same conditions as those in the antibacterial property test using LED light. Subsequently, 200 µL of the DMPO solution, in which a sample was immersed, was removed and the ROS were measured using an ESR spectrometer. 4-Hydroxy-2,2,6,6-tetramethylpiperidine-1-oxyl (TEMPOL, Sigma Aldrich, St. Louis, MO, USA) was used to quantify the hydroxyl radicals. A control ESR spectrum was obtained from a solution without sample immersion and visible-light irradiation. The amount of hydrogen peroxide (H_2O_2), which is an ROS, was measured using H_2O_2 colorimetry. Two types of solution were used for this purpose. Solution 1 was prepared by mixing 6 mL of 100 mM sulfuric acid and dissolving 11.8 mg of ammonium iron (II) sulfate hexahydrate into 30 mL of pure water. Solution 2 was prepared by dissolving 9.1 mg of xylenol orange tetrasodium salt and 2.186 g of sorbitol into 30 mL of pure water. A calibration curve was prepared using solutions 1 and 2, and 8.821 M H_2O_2 solution. A sample was placed in a 24-well plate and immersed in 500 µL of pure water. After irradiation with visible light for 30 min under the same conditions as those in the antibacterial property test using LED light, 400 µL of the pure water, in which the sample was immersed, was removed and poured into a glass tube. Subsequently, 200 µL of solution 1 and 200 µL of solution 2 were added into the glass tube and mixed well. The glass tube was then maintained at approximately 25 °C for 45 min. The absorbance of the mixture solution at a wavelength of 560 nm was then measured using ultraviolet-visible spectrophotometry (GeneQuant 1300, Biochrom, Ltd., Cambridge, UK).

4. Conclusions

TiO_2 co-doped with Cu and Ag was formed on the surface of Ti via NaOH-($Cu(NO_3)_2$ and $AgNO_3$) and heat treatments. The TiO_2 co-doped with Cu and Ag formed apatite on its surface in an SBF and showed higher antibacterial activity than that of TiO_2 doped with only Ag, especially under visible-light irradiation. The excellent antibacterial activity of TiO_2 co-doped with Cu and Ag under visible-light irradiation might be caused by the synergistic effect of the release of Ag and Cu and the generation of •OH from the sample. The toxicity of the sample needs to be evaluated in future studies, but dental implants with such a TiO_2 surface layer co-doped with Cu and Ag may reduce the risk of peri-implantitis.

Supplementary Materials: The following supporting information can be downloaded at: https://www.mdpi.com/article/10.3390/molecules28020650/s1, Figure S1: Electron spin resonance (ESR) spectra of control, AG, and AG-CU samples.

Author Contributions: Conceptualization, M.K.; methodology, K.S., T.M., M.I., M.F., K.Y., H.K. and M.S.; validation, K.S., T.M., H.K., M.S., T.Y. and M.K.; formal analysis, K.S., T.M., M.F., K.Y. and M.S.; investigation, K.S., T.M. and M.S.; data curation, K.S., T.M., M.F. and K.Y.; writing—original draft preparation, K.S.; writing—review and editing, M.K.; supervision, M.K.; project administration, M.K.; funding acquisition, M.K. All authors have read and agreed to the published version of the manuscript.

Funding: This research was funded by the Japan Society for the Promotion of Science KAKENHI [grant number: JP16H03177], the Kazuchika Okura Memorial Foundation, and the Institute of Biomaterials and Bioengineering, Tokyo Medical and Dental University [Project "Design and Engineering by Joint Inverse Innovation for Materials Architecture"] of the Ministry of Education, Culture, Sports, Science and Technology, Japan.

Institutional Review Board Statement: Not applicable.

Informed Consent Statement: Not applicable.

Data Availability Statement: The data will be available on request.

Conflicts of Interest: The authors declare no conflict of interest.

References

1. Atieh, M.A.; Alsabeeha, N.H.M.; Faggion, C.M.; Duncan, W.J. The frequency of peri-implant diseases: A systematic review and meta-analysis. *J. Periodontol.* **2013**, *84*, 1586–1598. [CrossRef] [PubMed]
2. Lee, C.T.; Huang, Y.W.; Zhu, L.; Weltman, R. Prevalences of peri-implantitis and peri-implant mucositis: Systematic review and meta-analysis. *J. Dent.* **2017**, *62*, 1–12. [CrossRef] [PubMed]
3. Dreyer, H.; Grischke, J.; Tiede, C.; Eberhard, J.; Schweitzer, A.; Toikkanen, S.E.; Glockner, S.; Krause, G.; Stiesch, M. Epidemiology and risk factors of peri-implantitis: A systematic review. *J. Periodont. Res.* **2018**, *53*, 657–681. [CrossRef] [PubMed]
4. Kotsakis, G.A.; Olmedo, D.G. Peri-implantitis is not periodontitis: Scientific discoveries shed light on microbiome-biomaterial interactions that may determine disease phenotype. *Periodontol. 2000* **2021**, *86*, 231–240. [CrossRef]
5. Salvi, G.E.; Cosgarea, R.; Sculean, A. Prevalence and mechanisms of peri-implant diseases. *J. Dent. Res.* **2017**, *96*, 31–37. [CrossRef]
6. Asensio, G.; Vizquez-Lasa, B.; Rojo, L. Achievements in the topographic design of commercial titanium dental implants: Towards anti-peri-implantitis surfaces. *J. Clin. Med.* **2019**, *8*, 1982. [CrossRef]
7. Shimabukuro, M. Antibacterial property and biocompatibility of silver, copper, and zinc in titanium dioxide layers incorporated by one-step micro-arc oxidation: A review. *Antibiotics* **2020**, *9*, 716. [CrossRef]
8. Dong, H.; Liu, H.; Zhou, N.; Li, Q.; Yang, G.W.; Chen, L.; Mou, Y.B. Surface modified techniques and emerging functional coating of dental implants. *Coatings* **2020**, *10*, 1012. [CrossRef]
9. Foster, H.A.; Ditta, I.B.; Varghese, S.; Steele, A. Photocatalytic disinfection using titanium dioxide: Spectrum and mechanism of antimicrobial activity. *Microbiol. Biotechnol.* **2011**, *90*, 1847–1868. [CrossRef]
10. Banerjee, S.; Dionysiou, D.D.; Pillai, S.C. Self-cleaning applications of TiO_2 by photo-induced hydrophilicity and photocatalysis. *Appl. Catal. B Environ.* **2011**, *176*, 396–428. [CrossRef]
11. Etacheri, V.; Di Valentin, C.; Schneider, J.; Bahnemann, D.; Pillai, S.C. Visible-light activation of TiO_2 photocatalysts: Advances in theory and experiments. *J. Photochem. Photobiol. C Photochem. Rev.* **2015**, *25*, 1–29. [CrossRef]
12. Suketa, N.; Sawase, T.; Kitaura, H.; Naito, M.; Baba, K.; Nakayama, K.; Wennerberg, A.; Atsuta, M. An antibacterial surface on dental implants, based on the photocatalytic bactericidal effect. *Clin. Implant Dent. Relat. Res.* **2005**, *7*, 105–111. [CrossRef]
13. Uchida, M.; Kim, H.-M.; Kokubo, T.; Fujibayashi, S.; Nakamura, T. Effect of water treatment on the apatite-forming ability of NaOH-treated titanium metal. *J. Biomed. Mater. Res.* **2002**, *63*, 522–530. [CrossRef] [PubMed]
14. Fujibayashi, S.; Nakamura, T.; Nishiguchi, S.; Tamura, L.; Uchida, M.; Kim, H.-M.; Kokubo, T. Bioactive titanium: Effect of sodium removal on the bone-bonding ability of bioactive titanium prepared by alkali and heat treatment. *J. Biomed. Mater. Res.* **2001**, *56*, 562–570. [CrossRef]
15. Sato, S. Photocatalytic activity of NO_x-doped TiO_2 in the visible light region. *Chem. Phys. Lett.* **1986**, *123*, 126–128. [CrossRef]
16. Sato, S.; Nakamura, R.; Abe, S. Visible-light sensitization of TiO_2 photocatalysts by wet-method N doping. *Appl. Catal. A-Gen.* **2005**, *284*, 131–137. [CrossRef]
17. Mathew, S.; Ganguly, P.; Rhatigan, S.; Kumaravel, V.; Byrne, C.; Hinder, S.J.; Bartlett, J.; Nolan, M.; Pillai, S.C. Cu-doped TiO_2: Visible light assisted photocatalytic antimicrobial activity. *Appl. Sci.* **2018**, *8*, 2067. [CrossRef]
18. Chen, M.; Wang, H.; Chen, X.; Wang, F.; Qin, X.; Zhang, C.; He, H. High-performance of Cu-TiO_2 for photocatalytic oxidation of formaldehyde under visible light and the mechanism study. *Chem. Eng. J.* **2020**, *390*, 124481. [CrossRef]
19. Janczarek, M.; Kowalska, E. On the origin of enhanced photocatalytic activity of copper-modified titania in the oxidative reaction systems. *Catalysts* **2017**, *7*, 317. [CrossRef]
20. Kaur, R.; Pal, B. Plasmonic coinage metal-TiO_2 hybrid nanocatalysts for highly efficient photocatalytic oxidation under sunlight irradiation. *New J. Chem.* **2015**, *39*, 5966–5976. [CrossRef]
21. Liu, M.; Sunada, K.; Hashimoto, K.; Miyauchi, M. Visible-light sensitive Cu(II)-TiO_2 with sustained anti-viral activity for efficient indoor environmental remediation. *J. Mater. Chem. A* **2015**, *3*, 17312–17319. [CrossRef]
22. Moniz, S.J.A.; Tang, J. Charge transfer and photocatalytic activity in CuO/TiO_2 nanoparticle heterojunctions synthesised through a rapid, one-pot, microwave solvothermal route. *Chem. Cat. Chem.* **2015**, *7*, 1659–1667.
23. Huang, L.; Peng, F.; Wang, H.; Yu, H.; Li, Z. Preparation and characterization of Cu_2O/TiO_2 nano–nano heterostructure photocatalysts. *Catal. Commun.* **2009**, *10*, 1839–1843. [CrossRef]

24. Liu, L.M.; Yang, W.Y.; Li, Q.; Gao, S.A.; Shang, J.K. Synthesis of Cu$_2$O nanospheres decorated with TiO$_2$ nanoislands, their enhanced photoactivity and stability under visible light illumination, and their post-illumination catalytic memory. *ACS Appl. Mater. Inter.* **2014**, *6*, 5629–5639. [CrossRef] [PubMed]
25. Park, D.; Choi, E.J.; Weon, K.-Y.; Lee, W.; Lee, S.H.; Choi, J.-S.; Park, G.H.; Lee, B.; Byun, M.R.; Baek, K.; et al. Non-invasive photodynamic therapy against-periodontitis-causing bacteria. *Sci. Rep.* **2019**, *9*, 8248. [CrossRef] [PubMed]
26. Kawashita, M.; Matsui, N.; Miyazaki, T.; Kanetaka, H. Effect of ammonia or nitric acid treatment on surface structure, in vitro apatite formation, and visible-light photocatalytic activity of bioactive titanium metal. *Colloid Surf. B Biointerfaces* **2013**, *111*, 503–508. [CrossRef]
27. Kawashita, M.; Yokohama, Y.; Cui, X.Y.; Miyazaki, T.; Kanetaka, H. In vitro apatite formation and visible-light photocatalytic activity of Ti metal subjected to chemical and thermal treatments. *Ceram. Int.* **2014**, *40*, 12629–12636. [CrossRef]
28. Kawashita, M.; Endo, E.; Watanabe, T.; Miyazaki, T.; Furuya, M.; Yokota, K.; Abiko, Y.; Kanetaka, H.; Takahashi, N. Formation of bioactive N-doped TiO$_2$ on Ti with visible light-induced antibacterial activity using NaOH, hot water, and subsequent ammonia atmospheric heat treatment. *Colloid Surf. B Biointerfaces* **2016**, *145*, 285–290. [CrossRef] [PubMed]
29. Iwatsu, M.; Kanetaka, H.; Mokudai, T.; Ogawa, T.; Kawashita, M.; Sasaki, K. Visible light-induced photocatalytic and antibacterial activity of N-doped TiO$_2$. *J. Biomed. Mater. Res. Part B* **2020**, *108*, 451–459. [CrossRef]
30. Suzuki, K.; Yokoi, T.; Iwatsu, M.; Mokudai, T.; Kanetaka, H.; Kawashita, M. Antibacterial properties of Cu-doped TiO$_2$ prepared by chemical and heat treatment of Ti metal. *J. Asian Ceram. Soc.* **2021**, *9*, 1448–1456. [CrossRef]
31. Kawashita, M.; Iwabuchi, Y.; Suzuki, K.; Furuya, M.; Yokota, K.; Kanetaka, H. Surface structure and in vitro apatite-forming ability of titanium doped with various metals. *Colloid Surf. A Physicochem. Eng. Asp.* **2018**, *555*, 558–564. [CrossRef]
32. Kizuki, T.; Matsushita, T.; Kokubo, T. Antibacterial and bioactive calcium titanate layers formed on Ti metal and its alloys. *J. Mater. Sci. Mater. Med.* **2014**, *25*, 1737–1746. [CrossRef] [PubMed]
33. Rajendran, A.; Pattanayak, D.K. Silver incorporated antibacterial, cell compatible and bioactive titania layer on Ti metal for biomedical applications. *RSC Adv.* **2014**, *4*, 61444–61455. [CrossRef]
34. Rajendran, A.; Pattanayak, D.K. Mechanistic studies of biomineralisation on silver incorporated anatase TiO$_2$. *Mater. Sci. Eng. C Mater. Biol. Appl.* **2020**, *109*, 110558. [CrossRef]
35. Hanaor, D.A.H.; Sorrell, C.C. Review of the anatase to rutile phase transformation. *J. Mater. Sci.* **2011**, *46*, 855–874. [CrossRef]
36. Ferraria, A.M.; Carapeto, A.P.; do Rego, A.M.B. X-ray photoelectron spectroscopy: Silver salts revisited. *Vacuum* **2012**, *86*, 1988–1991. [CrossRef]
37. Shimabukuro, M.; Manaka, T.; Tsutsumi, Y.; Nozaki, K.; Chen, P.; Ashida, M.; Nagai, A.; Hanawa, T. Corrosion behavior and bacterial viability on different surface states of copper. *Mater. Trans.* **2020**, *61*, 1143–1148. [CrossRef]
38. Takadama, H.; Kim, H.-M.; Kokubo, T.; Nakamura, T. TEM-EDX study of mechanism of bonelike apatite formation on bioactive titanium metal in simulated body fluid. *J. Biomed. Mater. Res.* **2001**, *57*, 441–448. [CrossRef]
39. Wang, X.-X.; Hayakawa, S.; Tsuru, K.; Osaka, A. Bioactive titania gel layers formed by chemical treatment of Ti substrate with a H$_2$O$_2$/HCl solution. *Biomaterials* **2002**, *23*, 1353–1357. [CrossRef]
40. Yang, B.; Uchida, M.; Kim, H.-M.; Zhang, X.; Kokubo, T. Preparation of bioactive titanium metal via anodic oxidation treatment. *Biomaterials* **2004**, *25*, 1003–1010. [CrossRef]
41. Wei, D.; Zhou, Y.; Jia, D.; Wang, Y. Characteristic and in vitro bioactivity of a microarc-oxidized TiO$_2$-based coating after chemical treatment. *Acta Biomater.* **2007**, *3*, 817–827. [CrossRef] [PubMed]
42. Uchida, M.; Kim, H.-M.; Kokubo, T.; Fujibayashi, S.; Nakamura, T. Structural dependence of apatite formation on titania gels in a simulated body fluid. *J. Biomed. Mater. Res. Part A* **2003**, *64*, 164–170. [CrossRef] [PubMed]
43. Kawashita, M.; Toda, S.; Kim, H.-M.; Kokubo, T.; Masuda, M. Preparation of antibacterial silver-doped glass microspheres. *J. Biomed. Mater. Res. Part A* **2003**, *66*, 266–274. [CrossRef] [PubMed]
44. Bajpai, S.K.; Bajpai, M.; Sharma, L. Copper nanoparticles loaded alginate-impregnated cotton fabric with antibacterial properties. *J. Appl. Polym. Sci.* **2012**, *126*, E319–E326. [CrossRef]
45. Imlay, J.A.; Linn, S. Bimodal pattern of killing of DNA-repair-defective or anoxically grown *Escherichia coli* by hydrogen peroxide. *J. Bacteriol.* **1986**, *166*, 519–527. [CrossRef]
46. Imlay, J.A.; Chin, S.M.; Linn, S. Toxic DNA damage by hydrogen peroxide through the Fenton reaction in vivo and in vitro. *Science* **1988**, *240*, 640–642. [CrossRef]
47. Linley, E.; Denyer, S.P.; McDonnell, G.; Simon, C.; Maillard, J.-Y. Use of hydrogen peroxide as a biocide: New consideration of its mechanisms of biocidal action. *J. Antimicrob. Chemother.* **2012**, *67*, 1589–1596. [CrossRef]
48. Takadama, H.; Kokubo, T. How useful is SBF in predicting in vivo bone bioactivity? *Biomaterials* **2006**, *27*, 2907–2915.

Disclaimer/Publisher's Note: The statements, opinions and data contained in all publications are solely those of the individual author(s) and contributor(s) and not of MDPI and/or the editor(s). MDPI and/or the editor(s) disclaim responsibility for any injury to people or property resulting from any ideas, methods, instructions or products referred to in the content.

Article

Electroanalysis of Ibuprofen and Its Interaction with Bovine Serum Albumin

Muhammad Dilshad [1], Afzal Shah [1,*] and Shamsa Munir [2]

[1] Department of Chemistry, Quaid-Azam University, Islamabad 45320, Pakistan
[2] School of Applied Sciences and Humanities, National University of Technology (NUTECH), Islamabad 44000, Pakistan
* Correspondence: afzals_qau@yahoo.com or afzalshah@qau.edu.pk

Abstract: The current work presents a sensitive, selective, cost-effective, and environmentally benign protocol for the detection of ibuprofen (IBP) by an electrochemical probe made of a glassy carbon electrode modified with Ag-ZnO and MWCNTs. Under optimized conditions, the designed sensing platform was found to sense IBP up to a 28 nM limit of detection. The interaction of IBP with bovine serum albumin (BSA) was investigated by differential pulse voltammetry. IBP–BSA binding parameters such as the binding constant and the stoichiometry of complexation were calculated. The results revealed that IBP and BSA form a single strong complex with a binding constant value of 8.7×10^{13}. To the best of our knowledge, this is the first example that reports not only IBP detection but also its BSA complexation.

Keywords: modified glassy carbon; ibuprofen; bovine serum albumin; inflammation

1. Introduction

Drugs facilitate the prevention and cure of diseases by strengthening the immune system. The mechanism of drug action is a specific biochemical interaction that results in targeted pharmacological effect. This action includes binding of the drug molecule to a specific targeted biological species such as enzymes or receptors [1–4]. Overdosage of drugs can result in adverse short-term or long-term health effects [5]. Drugs affect or alter the physiology of living organisms [6]. They stimulate a biological reaction by targeting macromolecules in the body [7]. As a rule, most drugs impede a particular biological response by interfering with the neurological system (particularly the brain). Based on pharmacodynamics, drugs can be classified as depressants, hallucinogens, and stimulants. Drugs can also be classified into analgesics and therapeutics. The current work presents electroanalysis of a non-steroidal anti-inflammatory drug (NSAID), ibuprofen (Scheme 1), which is the third most popular, prescribed, and sold-over-the-counter drug in the world [8]. The World Health Organization has listed ibuprofen (IBP) as an "essential drug" [9]. It is extensively used as a pain reliever in conditions such as menstrual cramps, headaches, arthritis, and a wide variety of other common aches and pains [10,11]. It plays an anti-inflammatory role by prohibiting the production of pro-inflammatory prostaglandins through inhibition of the enzyme cyclooxygenase [12].

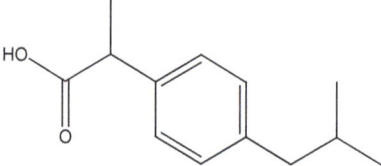

Scheme 1. Chemical structure of ibuprofen.

IBP enters into the environment due to improper pharmaceutical disposal during treatment. IBP manufacturing industries are a major contributor to the entrance of this drug into bodies of water. The sources of IBP water pollution can be seen in Figure S1. The toxicity of IBP metabolites exceeds that of the parent molecule. After excretion, IBP makes its way into wastewater treatment plants, sewage treatment plants, rivers, lakes, groundwater, soil, etc. A number of methods have been proposed for determining the concentration and effects of IBP in aquatic organisms [13]. To examine the short-term and long-term effects of IBP exposure, toxicity tests have been conducted on water-dwelling species. IBP is an emerging organic contaminant as its risk quotient is quite high [14]. Hence, part of the current work is focused on designing a sensitive electrochemical platform for the detection of IBP.

The protein–drug binding process involves complexation of a drug with protein. Protein–drug binding can be intracellular or extracellular. Intracellular binding involves the drug's interaction with cell proteins, eliciting a pharmacological response; the receptors with which the drug interacts are known as primary receptors. Extracellular binding does not usually result in a pharmacological response, and such drug receptors are known as secondary or silent receptors. The nature of a drug binding to a protein can be reversible or irreversible [15]. IBP pharmacokinetically interacts with BSA, which is a natural and very abundant (59%) plasma protein. BSA has a high affinity for binding with drug ligands and metabolites [16]. As shown in Figure 1, BSA has three domains (I, II, and III) and two sub-domains (A and B). The predicted drug-binding site is present in the sub-domains II A and III A.

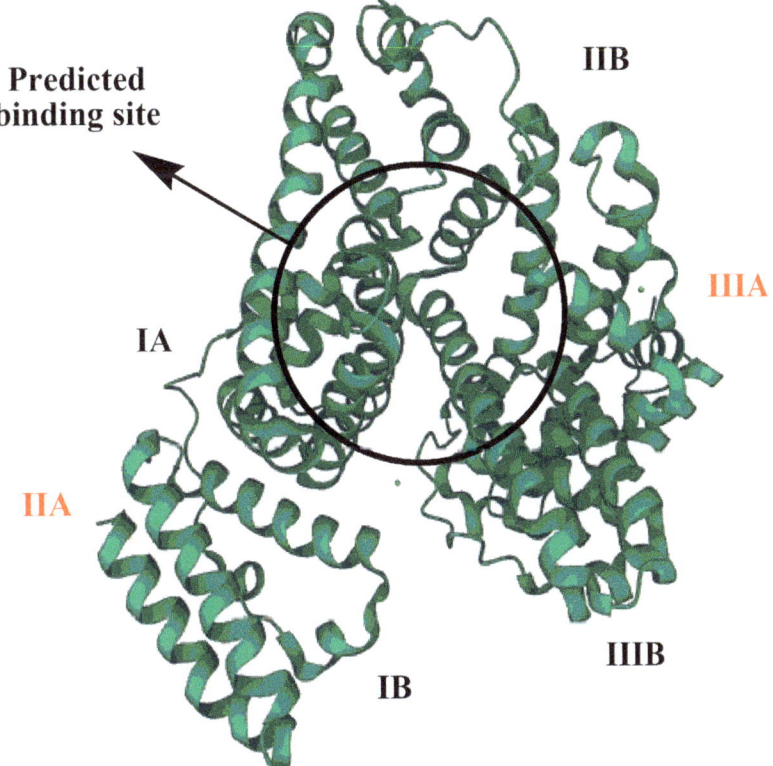

Figure 1. Bovine serum albumin (BSA).

Consider albumin (ALB) as the host protein that forms a complex with the guest drug (D). The binding constant (β_s), also known as the affinity constant or association constant, is determined from the stoichiometry of complexation between ALB and D according to the following equation [17]:

$$\text{ALB} + \text{mD} \overset{\beta_s}{\Leftrightarrow} \text{ALB} - \text{mD} \quad (1)$$

Here ALB − mD represents the protein drug complex, and the stoichiometric coefficient m indicates the number of drug molecules interacting per single molecule of protein. [FD] is the concentration of the free drug. The square brackets indicate the concentration of that particular species in Equation (2).

At equilibrium, β_s can be obtained by:

$$\beta_s = \frac{[\text{ALB} - \text{mD}]}{[\text{ALB}][\text{FD}]^m} \quad (2)$$

Electrochemical study shows that by adding albumin to a constant volume of drug solution, the peak current of the drug decreases owing to its interaction with the albumin. The maximum decrease in peak current occurs when the maximum amount of a drug interacts with the protein. The decrease in peak current occurs due to slow mobility of the drug–protein complex as compared to the free drug. Hence, involvement of more drug molecules in interaction with proteins results in a smaller amount of the free drug concentration in solution. Moreover, a greater concentration of albumin leads to a maximum decrease in peak current intensity due to the formation of a greater number of complex molecules. Mathematically:

$$\Delta I_{max} \propto [\text{ALB}]\text{total}$$

$$\Delta I_{max} = K\,[\text{ALB}]\text{total}$$

where K is the proportionality constant

$$\Delta I = K\,[\text{ALB} - \text{mD}] \quad (3)$$

$$\Delta I_{max} - \Delta I = K[[\text{ALB}]\text{total} - (\text{ALB} - \text{mD})] \quad (4)$$

As

$$[\text{ALB}]\text{total} = [\text{ALB}] + [\text{ALB} - \text{mD}]$$

By substituting the value of $[\text{ALB}]_{\text{total}}$, one obtains the following equation:

$$\Delta I_{max} - \Delta I = K[\text{ALB}] \quad (5)$$

$$\frac{\Delta I}{\Delta I_{max} - \Delta I} = \frac{[\text{ALB} - \text{mD}]}{[\text{ALB}]}$$

The right side of the above equation is equal to $\beta_s[\text{FD}]^m$ according to Equation (2). Hence,

$$\frac{\Delta I}{\Delta I_{max} - \Delta I} = \beta_s[\text{FD}]^m$$

$$\text{Log}\left(\frac{\Delta I}{\Delta I_{max} - \Delta I}\right) = \text{Log}\left(\beta_s[\text{FD}]^m\right)$$

$$\text{Log}\left(\frac{\Delta I}{\Delta I_{max} - \Delta I}\right) = \text{Log}\,\beta_s + m\,\text{Log}\,[\text{FD}] \quad (6)$$

A linear relationship between $\text{Log}\left(\frac{\Delta I}{\Delta I_{max} - \Delta I}\right)$ and Log [FD] indicates the formation of a single drug–protein complex. On the other hand, a non-linear relationship would suggest the formation of multiple complexes with different stoichiometry. If the $\Phi = \text{Log}\left(\frac{\Delta I}{\Delta I_{max} - \Delta I}\right)$

and Log [FD] plot shows two linear segments, then two types of complexes have been formed, and this will result in slopes m_1 and m_2. When multiple complexes are formed, then the following equation can be used:

$$\Delta I = \frac{\Delta I_1 \beta_1 [FD] + \Delta I_2 \beta_2 [FD]^2 + \cdots + \Delta I_n \beta_n [FD]^n}{1 + \beta_1 [FD] + \beta_2 [FD]^2 + \cdots + \beta_n [FD]^n} \quad (7)$$

where ΔI is the total decrease of peak current I_p obtained through a current voltage measurement.

$$f_1 = \frac{\Delta I}{[FD]}$$

$$f_1 = \frac{\Delta I_1 \beta_1 + \Delta I_2 \beta_2 [FD]^1 + \cdots + \Delta I_n \beta_n [FD]^{n-1}}{1 + \beta_1 [FD] + \beta_2 [FD]^2 + \cdots + \beta_n [FD]^n} \quad (8)$$

The binding constants obtained in the case of multiple types of complex formation can be discovered using detailed equations given in the Supporting Information File.

All the above mentioned equations for binding constant determination involve calculation from peak current of the drug in the presence of protein. In this regard, electroanalytical techniques are the most promising options. The importance of the detection of drugs and their metabolites using an electrochemical platform for safeguarding the health of patients has pushed researchers to develop sophisticated electroanalytical tools for their monitoring [18]. The detection method should possess the qualities of high selectivity and sensitivity so that it can be effective for sensing biotoxins [19]. In this regard, the current research work aims to design a modified glassy carbon electrode using a composite of multi-walled carbon nanotubes and Ag-doped ZnO (MWCNTs/Ag-ZnO) as an electrode modifier for the trace-level detection of IBP. Ag-ZnO was selected as it is an effective nanomaterial that exhibits electro-inactivity in the chosen potential window and its integration with MWCNTs results in significantly enhanced performance owing to the decrease in interfacial charge transfer resistance by producing a number of active sites for the adsorption of more analyte molecules and consequent intense current signal. Moreover, components of this modifier are environmentally benign.

2. Results and Discussion

2.1. Material Characterization

The synthesized Ag-ZnO nanoparticles were qualitatively characterized using X-ray diffraction analysis (XRD). The XRD diffractogram obtained for the synthesized nanoparticles is shown in Figure 2A. The peaks positioned at 31.75°, 34.37°, 36.15°, 47.55°, 56.48°, 62.81°, 66.37°, 67.90°, 69.12°, and 76.96° 2-theta values corresponded to the (100), (002), (101), (102), (110), (103), (200), (112), (201), and (004) diffraction planes, respectively. The X-ray diffractogram was in good agreement with the standard diffraction pattern of ZnO obtained from JCPDS card no. 36-1451.

SEM analysis of the surface characteristics of the produced Ag-ZnO revealed some interesting results. As synthesized, Ag-doped ZnO was visualized using an SEM micrograph (Figure 2B) with an accelerating voltage of 20 kV. Aggregation is indicated by the SEM micrograph, which reveals nanoparticles of varying sizes that have clumped together. The size can vary from one sample to the next depending on the condition of the precursors and the processes that were employed to synthesize the compound. The SEM shows the co-existence of smaller and larger nanoparticles. Agglomeration of smaller particles could be responsible for the formation of larger nanoparticles, and this phenomenon also explains why the forms of individual nanoparticles are obscured. The growing van der Waals forces, also known as intermolecular forces, between the silver nanoparticles and the zinc oxide network matrix caused the nanoparticles to become agglomerated. SEM studies indicated the particle size range at 11.7–20.8 nm, yet a large population of particles smaller than 11.7 nm is evident in the micrograph. The elemental composition of the synthesized material was analyzed using energy dispersive X-ray spectroscopy (EDX). The different elemental peaks such as Ag, Zn, O, S, and C were obtained as shown in Figure 2C. According to EDX analysis, the ZnO NPs with Ag doping were predominantly made up of Zn and O, with some trace amounts of Ag. Minor traces of S emerge from chemical impurities, whereas the presence of C originates from the carbon tape employed in the SEM analysis.

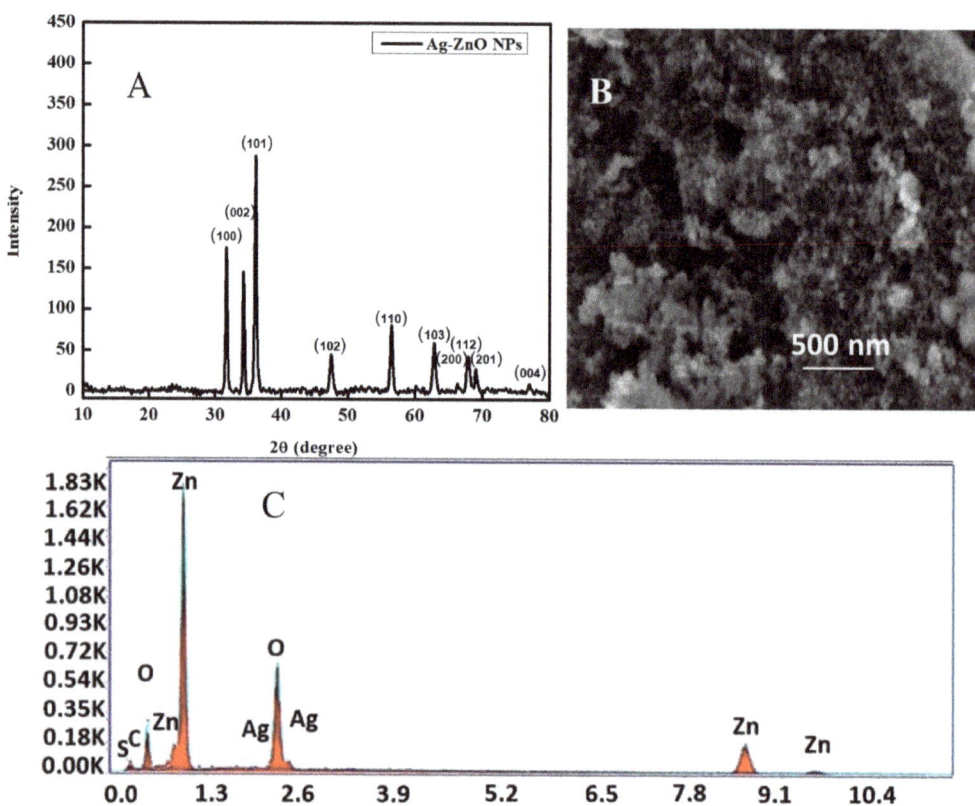

Figure 2. (**A**) X-ray diffractogram of Ag-ZnO, (**B**) SEM micrograph of Ag-ZnO, and (**C**) EDX spectrum of Ag-ZnO nanoparticles.

2.2. Electrochemical Characterization of IBP Using the Designed Sensor

The developed sensing platform was electrochemically characterized using cyclic voltammetry and electrochemical impedance spectroscopy. The surface area of the working electrode is a crucial aspect that has a considerable influence on the working ability of the electrochemical sensing platform. The cyclic voltammetric experiment was carried out to investigate the electroactive surface area of electrodes in a 5 mM solution of $K_3[Fe(CN)_6]$ (redox probe) in 0.1 M KCl (supporting electrolyte). The current response of $[Fe(CN)_6]^{3-/4-}$ for bare, Ag-ZnO-, MWCNT-, and MWCNT/ZnO-modified GCEs were investigated.

Figure 3A shows the cyclic voltammograms for 5 mM $[Fe(CN)_6]^{4-}$ obtained for the modified GCEs in 0.1 M KCl supporting electrolyte. In these voltammograms, the current peak at 0.34 V corresponds to the oxidation of $[Fe(CN)_6]^{4-}$ to $[Fe(CN)_6]^{3-}$, while the current peak at 0.21 V corresponds to the reduction of $[Fe(CN)_6]^{3-}$ to $[Fe(CN)_6]^{4-}$. From these voltammograms, the peak currents were found to be 100, 70, 48, and 25 µA for the MWCNT/Ag-AnO/GCE, MWCNT/GCE, Ag-ZnO/GCE, and bare GCE, respectively.

The current values were employed to calculate the electroactive surface area, and the Randles–Sevcik equation was utilized for both unmodified and modified electrodes [20].

$$I_p = 2.69 \times 10^5 \, n^{\frac{3}{2}} \, A D^{\frac{1}{2}} \, v^{\frac{1}{2}} \, C \tag{9}$$

where I_p represents the anodic peak current in amperes, D represents the diffusion coefficient of the analyte in cm^2 s^{-1}, A is the electroactive surface area in cm^2, v is the scan rate with a potential scan rate of V s^{-1}, C represents the concentration of the probe in mol cm^{-3}, and n represents the number of electrons.

For [K$_3$Fe(CN)$_6$], $D = 7.6 \times 10^6$ cm^{2}s^{-1} and $n = 1$. Table S1 demonstrates the electroactive surface areas of the GCE, Ag-ZnO/GCE, MWCNTs/GCE, and MWCNTs/Ag-ZnO/GCE. In comparison to the active surface area of the bare electrode (0.02 cm^2), the active surface area of the MWCNTs/Ag-ZnO/GCE (0.09 cm^2) was nearly 4.5 times greater. Figure 3A shows cyclic voltammograms for 5 mM [Fe(CN)$_6$]$^{3-}$ obtained for the modified GCEs in 0.1 M KCl supporting electrolyte. In these voltammograms, the redox couple corresponding to the oxidation of [Fe(CN)$_6$]$^{4-}$ to [Fe(CN)$_6$]$^{3-}$ and reduction of [Fe(CN)$_6$]$^{3-}$ to [Fe(CN)$_6$]$^{4-}$ are observable at 0.33 V and 0.17 V, respectively. An obvious twofold, threefold, and fourfold increase in peak current can be seen for the Ag-ZnO/GCE, MWCNTs/GCE, and MWCNTs/Ag-ZnO/GCE, respectively. The faster electron transport of the redox probe can be related to better conductivity and more active sites provided by the modifier components at the GCE surface.

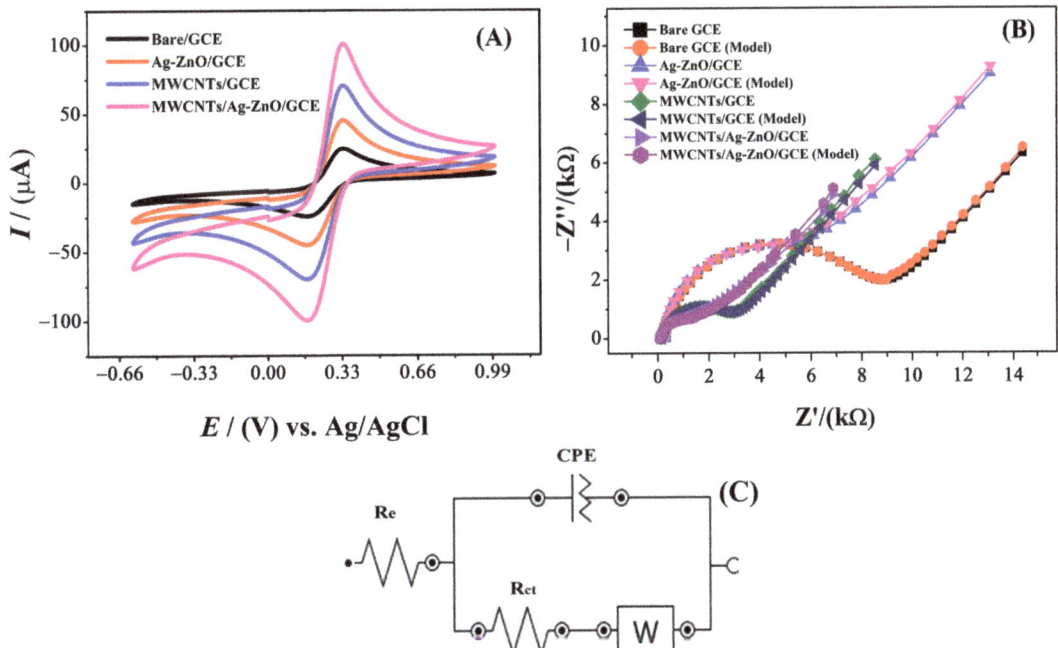

Figure 3. (**A**) Cyclic voltammograms for bare and modified glassy carbon electrodes recorded in 5 mM K$_3$[Fe(CN)$_6$] with 0.1 M KCl as supporting electrolyte at a scan rate of 100 mVs^{-1} (**B**) Nyquist plots of bare and modified GCEs in a solution of 5 mM K$_3$[Fe (CN)$_6$] as a redox probe and 0.1 M KCl as a supporting electrolyte, and (**C**) equivalent circuit corresponding to EIS data.

The method of electrochemical impedance spectroscopy did not impart any damage to the tested material. Through EIS, charge transfer kinetics for both bare and modified GCEs were investigated in a 5 mM solution of $K_3[Fe(CN)_6]$ in a 0.1 M KCl solution [21]. The DC potential at 0 V and 10 mV was set as the peak-to-peak amplitude of the AC potentials superimposed on the aforementioned DC potential. Figure 3B depicts the Nyquist plots that were produced at frequencies ranging from 100 kHz to 0.1 Hz with an amplitude of 10 mV for the bare GCE, Ag-ZnO/GCE, MWCNTs/GCE, and MWCNTs/Ag-ZnO/GCE. The diameter of the semicircle in the Nyquist plot between the imaginary impedance (Z'') versus real impedance (Z') represents the resistance to charge transfer (R_{ct}), while the linear part in the lower frequency region arises from diffusion limited processes characterized by Warburg impedance (W_d) [22,23]. The semicircular section at a greater frequency corresponds to charge transfer resistance. A semicircle of smaller diameter represents lower R_{ct}, and vice versa [24]. Table S2 shows R_{ct} values of 8173, 4277, 2627 and 1610 Ω for bare, Ag-ZnO-, MWCNTs- and MWCNTs/Ag-ZnO-modified GCEs, respectively. It can be seen that R_{ct} values decreased for modified electrodes as the surface area of the electrodes increased. The increase in the surface area of the electrode owing to adsorbed molecules of the MWCNTs/Ag-ZnO/GCE and its mediator role between the transducer GCE and the redox probe caused the reduction in the R_{ct} value. The availability of the active sites on the GCE increased due to the presence of the Ag-ZnO and MWCNT molecules. Immobilized molecules on the GCE surface link the analyte molecules to the electrode [25]. The modifier molecules promote ease of electron transfer between the analyte and the electrode. In addition, Ag-ZnO and MWCNTs decrease interfacial charge transfer resistance. Therefore, the selected modifier renders the surface of the GCE an excellent sensing substrate for detecting the analyte. The EIS-derived parameters in Table S2 suggest that the impedance parameters have changed because of electrode modification. This indicates that the electrode surface was successfully fabricated. By modifying the electrode with the MWCNTs/Ag-ZnO/GCE, there is an increase in the electron transfer rate between the analyte and the electrode and it decreases the charge transfer resistance. The CV and EIS data are in good agreement. In CV, the MWCNTs/Ag-ZnO/GCE shows maximum current, and EIS shows minimum resistance. Therefore, the MWCNTs/Ag-ZnO/GCE was chosen as a reliable electroanalytical sensing platform for the detection of the analyte IBP.

2.3. Voltammetric Analysis of the Targeted Analyte

The peak current response for IBP in 0.9 M NaOH with bare and modified electrodes was studied using DPV in a potential window ranging from 0.4 V−1.7 V, as illustrated in Figure 4. The single oxidation peak depicts the oxidation of the −OH group in the IBP molecule. The modified electrode, MWCNTs/Ag-ZnO/GCE, exhibits the highest current response, with an approximately two times greater current intensity as compared to the signal for the bare GCE. The modified electrode (MWCNTs/Ag-ZnO/GCE) possesses greater catalytic performance due to the role of Ag-ZnO and MWCNTs in accelerating electron transport between the transducer and the analyte. Both MWCNTs and Ag-ZnO enhance the surface area by providing a greater number of electroactive sites to which a greater number of the analyte molecules (IBP) could be anchored. The consequent closer accessibility of IBP to the transducer led to peak current intensification at the modified electrode surface as compared to the unmodified electrode. Hence, owing to a greater number of IBP molecules on the modified interface, more molecules can be oxidized at the given potential, resulting in a greater response in terms of anodic current.

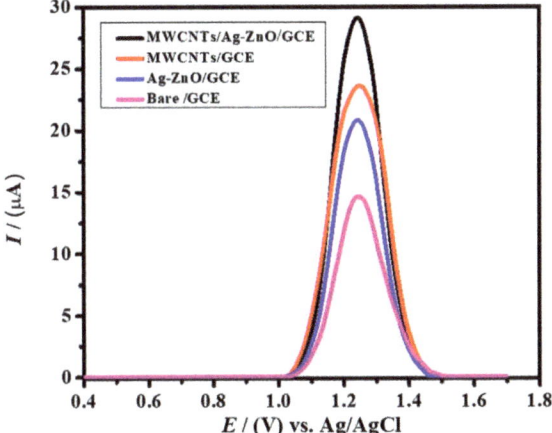

Figure 4. DPVs of 0.09 mM ibuprofen on bare GCE, Ag-ZnO/GCE, MWCNTS/GCE, and MWCNTs/Ag-ZnO/GCE in a supporting electrolyte of 0.9 M NaOH.

2.4. Effect of Scan Rate

To analyze the effect of scan rate on the oxidation peak current of IBP, cyclic voltammograms were obtained using the MWCNTs/Ag-ZnO/GCE. At scan rates ranging from 50 to 350 mV s^{-1}, cyclic voltammograms were recorded to observe the nature of the redox reaction, i.e., whether it was a diffusion or surface-controlled electrochemical process. As the scan rate increased, the peak current intensity increased proportionally (see Figure 5).

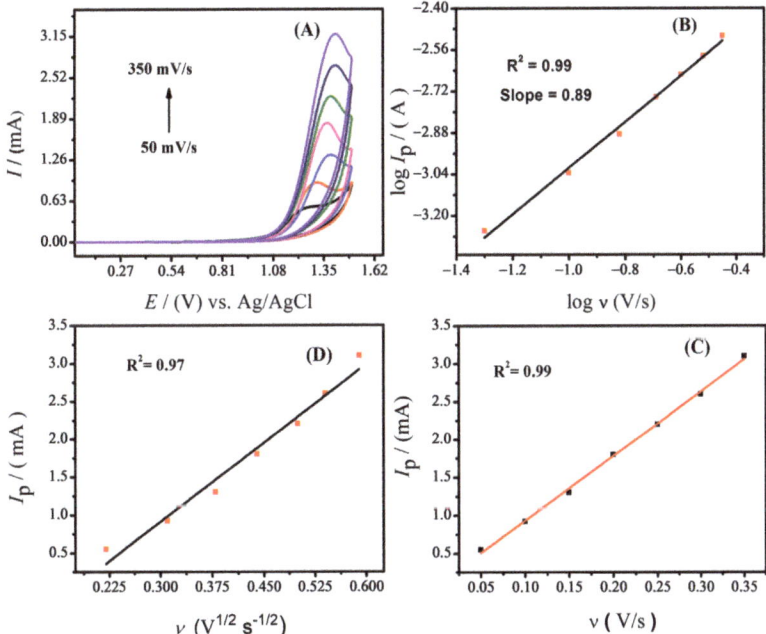

Figure 5. (**A**) Effect of different scan rates on the anodic peak current of IBP in supporting electrolyte (0.9 M NaOH), (**B**) calibration plot of IBP between log I_p vs. log v, (**C**) plot of peak current vs. scan rates of IBP oxidation, and (**D**) calibration plot of IBP between peak current vs. square root of scan rates.

The slope value of the log of I_p and log of v plot can be used to deduce the nature of the redox process. If the value of the slope is 1, the process should be controlled by adsorption; on the other hand, if the slope value is 0.5 the process should be controlled by diffusion [26]. The slope value of 0.89 as depicted in Figure 5B suggests that the electrochemical oxidation of IBP is controlled by both processes [27]. The straight-line equation for the graph shown in Figure 5B is represented by log I_p = 0.89 and log v − 6.57. Since the correlation coefficient in the plot of oxidation peak current vs. scan rate is higher (Figure 5C) than that of I_p vs. $v^{\frac{1}{2}}$ (Figure 5D), the process of adsorption works better on the electrode's surface.

2.5. Optimization of Experimental Parameters

Among all the voltammetric techniques, DPV is the most sensitive pulse technique. Its detection limit is comparable to that achieved by chromatographic and spectroscopic approaches. After the sensing ability of the modified GCEs was tested with EIS, CV, and DPV, and the maximum current response was measured, the DPV method was used to find the best combination of experimental parameters to obtain the maximum current response. The following sections provide more information on these parameters.

2.5.1. Supporting Electrolyte Optimization

Supporting electrolytes reduce the electrical potential gradient, which removes migration in a signal and minimizes the ohmic drop effect. Alternately, it increases the conductance of the solution. Supporting electrolytes impact the peak shape, position (potential), and intensity. The analyte (IBP) was tested in various supporting electrolytes to obtain optimized results. Different electrolytes were used, such as CH_3COOH, H_2SO_4, NaCl, NaOH, KCl, acetate buffer (pH = 7), BRB (pH = 7), PBS (pH = 7), and KOH. The highest current response (30 µA) and the clearest definition of peak form was noticed in the solution containing sodium hydroxide as the supporting electrolyte in comparison to other electrolytes, as can be seen in Figure S2A. The high electrical conductance in NaOH may be attributed to its high solubility in water as compared to the other electrolytes investigated in this work. Moreover, it does not produce any gases and as such its concentration remains constant over time. Therefore, NaOH was selected for further electrochemical investigations.

2.5.2. Effect of Accumulation Potential

An optimized accumulation potential with a value lower than the analyte's oxidation potential during the potential sweep helps in accumulation of most of the analyte's molecules on the surface of the electrode. This results in intense oxidation signal generation during anodic potential scanning. Hence, the impact of accumulation potential on the oxidation peak current for IBP was assessed using anodic stripping DPV. The deposition potential was in the range of −1.6 to 0.3 V. The peak current of the analyte was enhanced with the increment of deposition potential up to −1.2 V as depicted in Figure S3A. Therefore, −1.2 V deposition potential was selected for further electrochemical investigations of IBP (see Figure S3B). It is speculated that all available active sites become saturated with IBP molecules at the accumulation potential of −1.2 V and a further increase in the accumulation potential may disturb the proper orientation of IBP molecules at the electrode–electrolyte interface, thus resulting in a decrease in the current response.

2.5.3. Influence of Accumulation Time

The performance of the designed electrochemical scaffold can be improved by optimizing the deposition time to obtain an enhanced peak current. Under optimized deposition potential conditions, accumulation time was varied in the range of 5 s to 50 s. IBP displayed a maximum current intensity at an accumulation time of 30 s (see Figure S4A). As the number of accessible active sites on an electrode surface increases, the peak current intensity continues to rise with an increase in the amount of time spent in depositing material. A greater concentration of analyte is accumulated on its surface. The largest peak current is seen at the point of saturation, which occurs when the analyte has become oriented to all of the available active sites. The molecules of the analyte need to be aligned in the right direction for the best possible deposition. The reduction of drug molecules occurs when they are accumulated on the electrode surface. These molecules undergo an oxidation process in the anodic stripping differential pulse voltammograms when the potential is varied from negative to positive values. At a deposition time of 30 s, maximum peak current is observed for IBP. The peak current intensity of the analyte is negatively influenced by further increasing the deposition time, as evidenced in Figure S4B.

2.6. Limit of Detection of IBP and Calibration Plot

DPV was performed to examine the LOD of IBP under optimized conditions, i.e., 0.9 M NaOH, -1.2 V accumulation potential, and 30 s deposition time. Figure 6A demonstrates that the concentration of the analyte influences the peak current. Using DPV, analyte solutions of different concentrations were tested to locate the sensor's absolute minimum sensitivity. Figure 6B shows the electrochemical current response of the analyte at a variety of concentrations. The linear calibration curve was obtained between 0.1 to 90 μM analyte concentration. The limit of detection was calculated using the formula 3 σ/m, where m is the slope of the plot of peak current versus concentration and σ is the standard deviation of the blank signal. To compute the standard deviation, currents of the blank solution at the peak point were utilized. The LOD for the designed sensor MWCNTs/Ag-ZnO/GCE was found to be 28 nM for IBP.

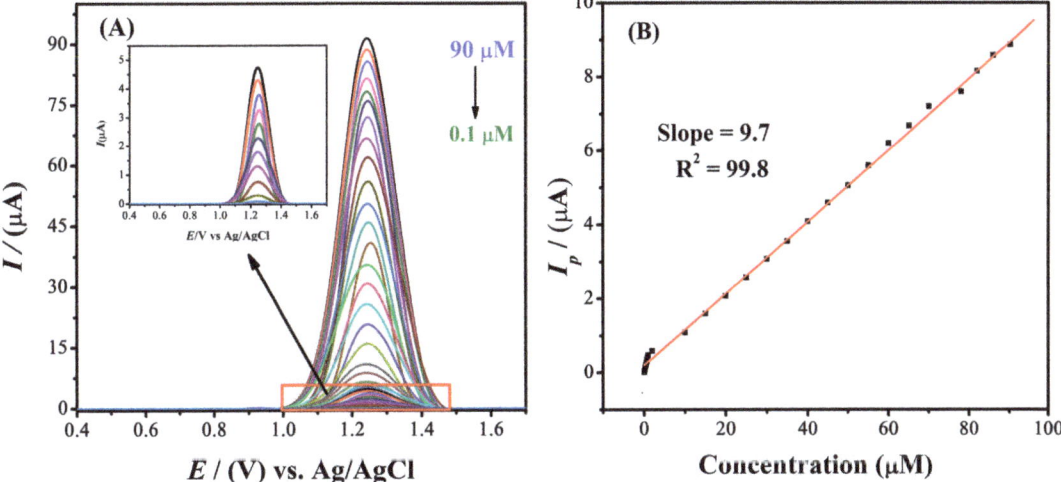

Figure 6. (**A**) Differential pulse voltammograms for different concentrations of IBP using an MWCNTs/Ag-ZnO/GCE under pre-optimized conditions at a scan rate of 10 mVs^{-1}, (**B**) calibration plot obtained by DPV data for various concentrations of IBP.

One particularly interesting aspect of the developed sensor is its ability to function over a wide linearity range, from about 0.1 μM to around 90 μM (Figure 6B). It can be seen from Table 1 that the LOD value of 28 nM is significantly lower than the previously reported data for the different designed sensors. Therefore, it can be concluded that the modified electrode is a promising platform for the analytical detection of IBP.

Table 1. Comparison of design sensors with reported sensors.

Sensors	Measurement Technique	Linear Range (µM)	LOD (nM)	Ref.
Pretreated GCE	SWSV	1.45–3.87	960	[28]
SD-MWCNT/GCE	FIA-AMP	10–1000	1900	[29]
Polyaniline nanofiber/GCE	DPSV	0.96–1.94	480	[30]
P(L-Asp)/GCE	SWV	1–150	220	[31]
MWCNT–CPE	DPV	2.36–242	9100	[32]
Clay-CPE	DPV	1–1000	835	[33]
HKUST-CNF	CV	4.84–29.08	100	[34]
Pd-PdO/Mt-CPE	DPV	0.01–0.9	28	[35]
AgNPs@Af-GO-MIP/GCE	DPV	1–100	8.7	[36]
MWCNTs/Ag-ZnO/GCE	DPV	0.1–90	28	This work

2.6.1. Estimation of the Stability of the Designed Sensor

The repeatability and reproducibility of the established sensing platform were used to evaluate its stability and practical applicability. The electrochemical response of the sensor in the presence of IBP under pre-optimized testing conditions was used to evaluate the sensor stability. DPV analysis was performed on the MWCNTs/Ag-ZnO/GCE after it had been placed in NaOH for various amounts of time to assess its stability. Peak current intensity did not significantly vary (< %) with signal intensity on the newly modified electrode up to 36 h, as shown in Figure 7A. The sensing platform showed intra-day and inter-day stability of response in terms of current, remaining similar up to 36 h. Due to the poor water solubility of the components of electrode modifiers, the developed sensor showed stability over a range of time intervals. This not only prevented the modifier from eroding from the electrode surface but also kept the peak current of the analyte stable over time. Four separate MWCNTs/Ag-ZnO/GCEs were prepared and then subjected to DPV analysis for a reproducibility check of the designed sensor. The results displayed in terms of the oxidation peak show no significant deviation, asserting outstanding repeatability and reproducibility, as illustrated in Figure 7B. The percent RSD reproducibility of 0.67 and percent RSD repeatability of 1.05 are influential figures of merit for the designed sensing platform.

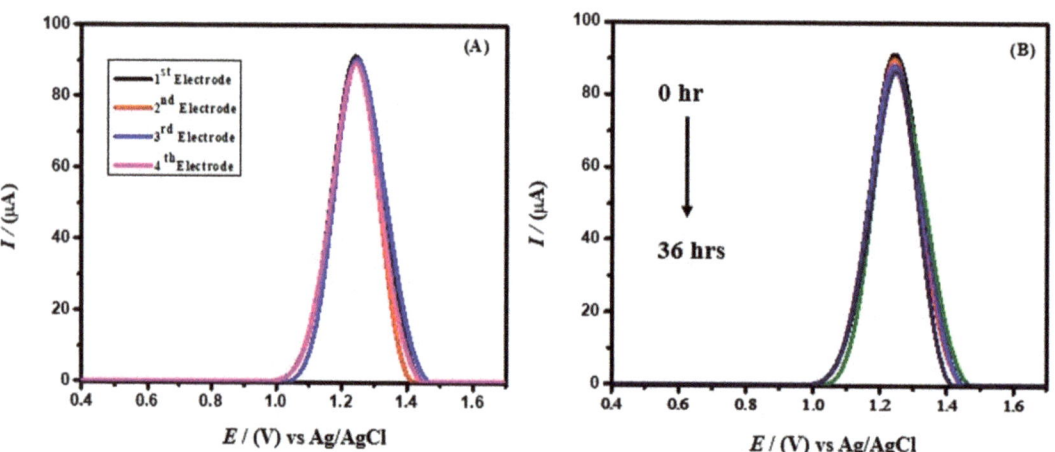

Figure 7. (**A**) DPVs of IBP showing reproducibility of the designed sensor using supporting electrolyte NaOH (**B**) DPVs of IBP for the designed sensor at different time intervals.

2.6.2. Effects of Interferents for Validation of the Designed Sensor

A real sample collected from a living being or a waste disposal site of a pharmaceutical industry or hospital may be composed of chemical species other than ibuprofen that may serve as potential interferents in the detection of the said analyte. To mimic the potential effect of the interferents, a number of chemical species, including metal ions and essential textile dyes, were individually spiked at 1 mM concentration. The voltammetric responses in the presence of interferents suggest that the IBP current signal at the designed sensor is not significantly influenced, which shows that the designed platform possesses discrimination ability for the target analyte (Figure 8).

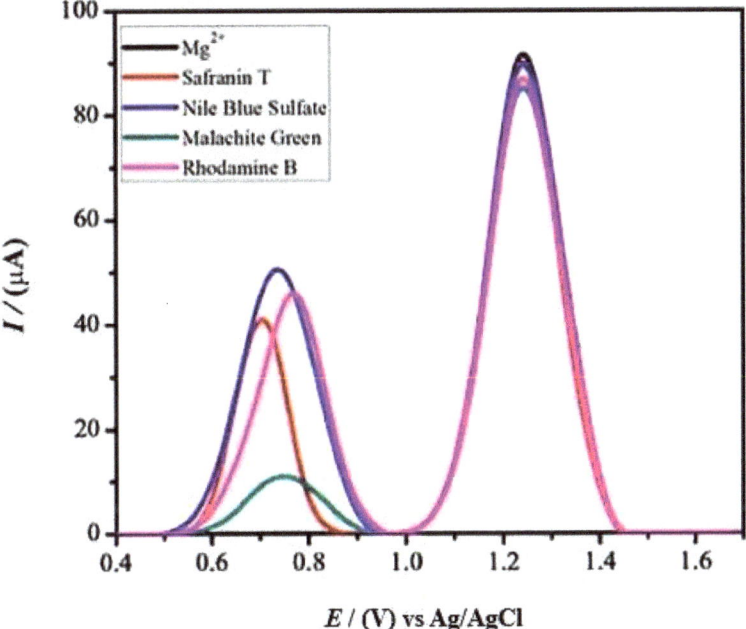

Figure 8. Differential pulse voltammograms for 90 µM IBP in the presence of 1 mM different interfering agents.

2.7. Interaction Studies of IBP with BSA

Prostaglandins (PGs) are a group of lipids produced in areas of tissue damage and infection and are associated with the sensation of pain, fever, and inflammation. IBP lowers the level of PG in the body by inhibiting the cyclooxygenase enzyme, which is required for the synthesis of PG and hence can reduce pain and inflammation by lowering the PG level. Considering this effect of IBP, we evaluated the binding of IBP with bovine serum albumin (BSA, an enzyme required for synthesis of PG) using DPV. Traditionally, equilibrium dialysis has been used to evaluate the binding constants of a drug to plasma proteins. However, this method has several shortcomings, including lengthy equilibration times, usually 12–48 h. The requirement for initial studies to determine the time in which the system attains equilibrium has prompted the researchers to develop alternate techniques [37,38].

For interaction studies of IBP with BSA, DPV is superior to linear scan voltammetry because of its sensitivity and ability to reduce the comparatively high background currents caused by the presence of albumin in solution [39]. For binding constant determination, all the values for peak current and its punctual difference, free drug concentration, binding constant, stoichiometry, etc., were repeated multiple times and reported values represent their means (with RSD±10% of the given values). DPV was carried out for various concentrations of IBP in the presence of 0.9 M NaOH in a potential window ranging from 0.4 V to 1.7 V at a scan rate of 10 mV s^{-1} and step potential of 5 mV (Figure 9A). IBP concentration was varied, and with decreasing concentrations peak current decreased as expected. The voltammograms were first obtained in the absence of BSA. Then, BSA in large excess (1 mM) was added to a solution of the drug, along with 0.9 M NaOH as supporting electrolyte. Figure 9B depicts the voltammograms obtained in the presence of BSA. The peak current was significantly decreased compared to the peak current of voltammograms obtained in the absence of BSA. For example, at 0.23 µM IBP, the addition of BSA decreased the current from 0.55 µA to 0.09 µA. The decreasing peak current indicates interaction between IBP and BSA, leaving less free IBP in solution for electrochemical oxidation at the electrode. A maximum decrease in peak current occurs when the maximum amount of the drug reacts with protein. Hence, the decrease in peak current is attributed to the interaction between the drug and BSA.

From the differential pulse, voltammograms recorded in Figure 9A,B show current values in the absence and presence of BSA; their punctual differences were calculated as listed in Table S3.

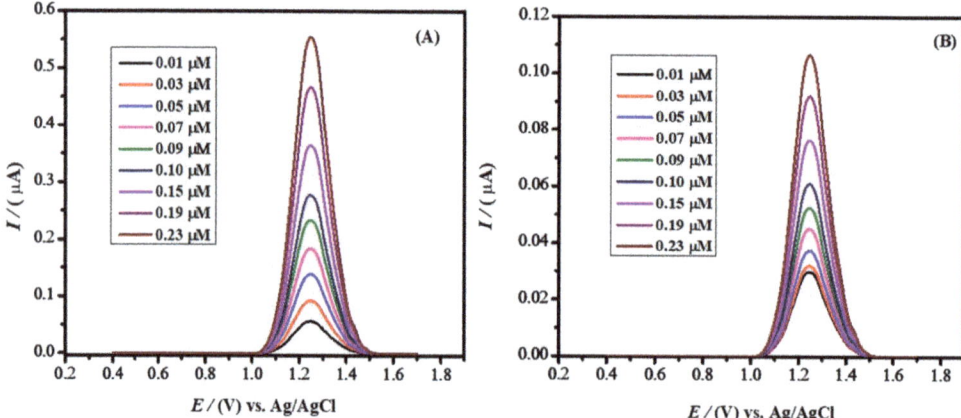

Figure 9. (**A**) DPV of IBP recorded with a supporting electrolyte of 0.9 M NaOH at different concentrations in the absence of BSA (**B**) DPV of IBP recorded with a supporting electrolyte of 0.9 M NaOH at different concentrations in the presence of 1 mM BSA.

The voltammetric calibration curve of IBP in 0.9 M NaOH was registered with IBP concentrations (C_{drug}) in the range of 0.01 to 0.23 µM (Figure 10A). The resulting linear plot I_{TD} vs. C_{drug} was obtained, where I_{TD} represents current due to the total amount of the drug (see Figure 10A). In the same way, a voltammetric calibration plot with the same amounts of the drug (C_{drug}) and a fixed amount of albumin was recorded ([ALB]total), operated with an albumin concentration of 1 mM. The resulting linear plot I_{FD} vs. C_{drug} was obtained, where I_{FD} represents the current of the free drug. For any point on the calibration curve, we may calculate the difference between the two values at that point ($\Delta I = I_{TD} - I_{FD}$), because the complex is not electroactive under operating conditions.

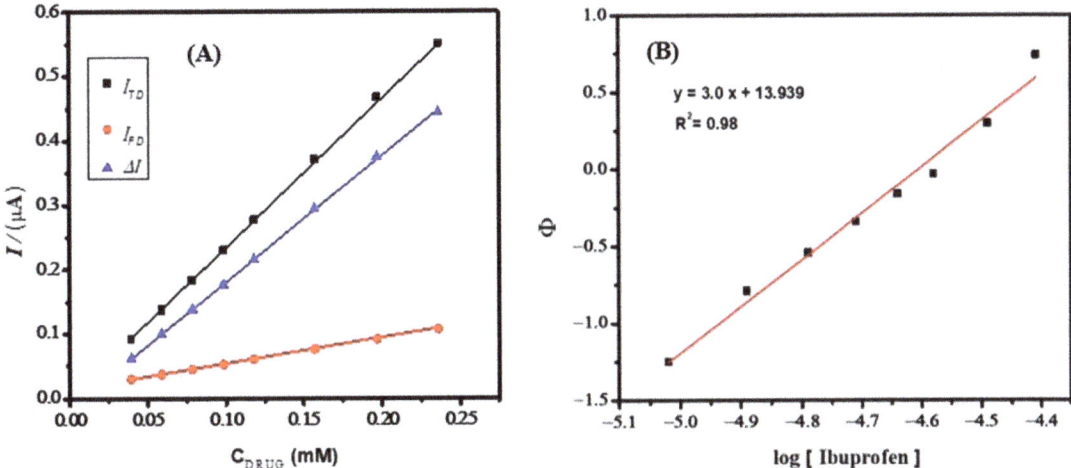

Figure 10. (**A**) Relationship of peak current vs. drug concentration for IBP and no BSA, NaOH 0.9 M, and 5 μM BSA + NaOH 0.9 M, and punctual difference between the values of I_{TD} and I_{FD} ($\Delta I = I_{TD} - I_{FD}$) (**B**) relationship between Φ and log [FD] in the case of IBP with BSA, an example of a formation with a single type of complex.

The concentration of free drug [FD] was evaluated by considering the calibration plot obtained in the absence of BSA with data given in Table S4. By plotting all values of log [FD] vs Φ, a linear plot was obtained (Figure 10B) according to equation 6 mentioned in the introduction. The value of m determined from the slope shows the number of drug molecules interacting per single molecule of BSA. The binding constant ($β_s$) with a value of 8.7×10^{13} was evaluated from the antilog of the intercept. A comparison of the binding constant of IBP–BSA and its stoichiometry with reported values is given in Table 2. A reasonably stronger binding of IBP with BSA is required to inhibit the functioning of BSA. The comparison of binding constant values for various protein–drug complexes shows that IBP–BSA has the highest value of binding constant, thus indicating effective inhibition of BSA by IBP.

Table 2. Characteristics of drug complexation with OVA, BSA, and HAS.

Drug Complexes	Complex Stoichiometry (m)	Binding Constant	Ref.
Ketoprofen-BSA	3	2.4×10^9	[40]
Ketoprofen-HAS	1	1.4×10^{10}	[41]
Lorazepam-OVA	3	2.5×10^{10}	[42]
Paroxetine-BSA	4	5.8×10^{18}	[43]
Paroxetine-OVA	3	2.6×10^{23}	[44]
Ibuprofen-BSA	3	8.7×10^{13}	This work

3. Conclusions

A quick-responding, sensitive, and stable electrochemical sensor was developed using a composite of MWCNTs and Ag-doped ZnO nanoparticles as a modifier of a GCE surface. The designed sensor demonstrated excellent ability to detect IBP down to 28 nM. The peak current response of IBP was greatly improved by the components of the recognition layer in comparison to the bare GCE. Cyclic voltammetric and electrochemical impedance spectroscopic investigations revealed that the designed sensing scaffold has 4.5 times greater electroactive surface area and approximately 5 times less interfacial charge transfer resistance as compared to the bare GCE, which results in the generation of an intense signal for IBP oxidation. Voltammetric analysis revealed that the modified electrode possesses inter-day durability and four individually fabricated electrodes exhibited unaltered efficiency in terms of repeatability. The PG inhibition behavior of IBP was also investigated by measuring its binding capacity with BSA using DPV, where it was revealed that three molecules of IBP bind to a single molecule of BSA with a binding constant value of 8.7×10^{13}. This strong binding potency of IBP suggests that it can play a significant role in PG inhibition and in turn can be developed as a medicine for reducing pain and inflammation. Considering the importance of IBP in the pharmaceutical industry and its medical potential, new and innovative analytical methods with high efficiency are still needed to effectively control this non-steroidal anti-inflammatory drug in pharmaceutical doses and to detect it in biological fluids. Moreover, coupling such sensors with industries for the early sensing of IBP in industrial effluents before their release to freshwater bodies may broaden their future applicability.

4. Experimental Section

4.1. Materials and Methods

IBP was obtained from a bio-lab pharmaceutical company (Islamabad, Pakistan) and was used as received. BSA and MWCNTs (purity > 95%) were obtained from Sigma Aldrich. CH_3COOH, H_2SO_4, NaCl, NaOH, KCl, acetate buffer, Britton–Robinson buffer (BRB), phosphate-buffered saline (PBS), and KOH were tested as supporting electrolytes. Zinc acetate dihydrate and silver nitrate were purchased from Sigma Aldrich. PBS was prepared by dissolving a specified amount of Na_2HPO_4 and NaH_2PO_4 in distilled water and using 0.1 M HCl and 0.1 M NaOH for pH adjustment.

The nanomaterials were characterized using X-ray diffraction spectroscopy (X-ray diffractometer model Analytical 30440/60 X per PRO with copper K_α radiation source, scan rate of 0.01, and 2θ range from 10°–80°) and scanning electron microscopy (JEOL.JAD-2300 module, Tokyo, Japan). A Metrohm multichannel Autolab (M101, PGSTAT302N, Utrecht, The Netherlands) equipped with NOVA 1.11 software and Gamry Interface 5000E potentiostat were used for electrochemical measurements. The electrochemical cell consisted of a glass cell with two layers of glass walls and a Teflon cover. The cap had five standard taper ports; three of them were used for the introduction of electrodes (the Ag/AgCl reference electrode, the working electrode, and the Pt auxiliary electrode), while the other two were used for the entrance and exit of inert gas purging.

4.2. Synthesis of Ag-ZnO

A hydrothermal method was employed to synthesize ZnO nanoparticles. Zinc acetate dihydrate (0.473 g) was dissolved in 25 mL ethanol and the pH of the solution was adjusted by dropwise addition of 1 M solution of NaOH. The solution was stirred for 10 min and then transferred to an autoclave and placed in the oven at a temperature of 200 °C for 6 h. The product was filtered and washed several times with ethanol to neutralize the pH, followed by drying at 80 °C. Ag-doped ZnO was prepared by following the same procedure except that a known amount of Ag precursor was added to the solution along with the zinc acetate precursor.

4.3. Electrode Modification and Detection Procedure

A clean glassy carbon electrode surface was obtained by rubbing it on a pad with 0.5 μm alumina slurry in a figure-eight pattern to keep the surface even. The surface was rinsed with a stream of distilled water to get rid of any unwanted particles. This process produced an impurity-free cleaned surface with a silver mirror-like finish. The cleanliness of the GCE was ensured by obtaining cyclic voltammograms in a potential window ranging from 0.4 V to 1.7 V that reflected the reproducibility of the obtained voltammograms [19].

The stock solution of IBP was prepared in a 1:1 mixture of distilled water and ethanol. First, a 5 μL droplet of MWCNTs with a concentration of 1 mg/mL was drop-casted on two separate pre-cleaned GCEs, followed by drop-casting 10 μL of Ag-ZnO; they were then subjected to drying in a vacuum oven at 50 °C. The performance of the designed sensing platform (MWCNTs/Ag-ZnO/GCE) was examined using differential pulse voltammetry (DPV) for the detection of IBP. The DPV was carried out at a step potential of 5 mV and scan rate of 10 mV/s. Electrochemical impedance spectroscopy (EIS) was employed to obtain the impedimetric results at an amplitude of 10 mV in the frequency range of 100 kHz to 0.1 Hz. Gamry software version 7.05 was used to fit an equivalent circuit to the obtained data and the results for the modified electrode were then compared with those of the bare electrode. Different experimental parameters such as deposition time, deposition potential, and pH of the medium were optimized, and the limit of detection (LOD) of IBP was obtained under optimized conditions. For the IBP–protein binding studies, varying concentrations of IBP were added to a 0.9 M solution of NaOH in the presence of an excess concentration of BSA (1 mM). The decrease in peak current of IBP was used for the quantification of the IBP−BSA binding constant.

Supplementary Materials: The following supporting information can be downloaded at: https://www.mdpi.com/article/10.3390/molecules28010049/s1, Figure S1: Source of ibuprofen pollution.; Table S1: Electroactive surface areas of bare and modified electrodes; Table S2: EIS-derived parameters. Solution resistance (R_s), Charge transfer resistance (R_{ct}), constant phase element (CPE). Figure S2: (A) Effect of various supporting electrolytes on the anodic peak current of IBP using MWCNTs/Ag-ZnO modified GCE (B) Bar graph of IBP between peak current vs. various supporting electrolytes; Figure S3: (A) Effect of deposition potential on the peak current of 0.09 mM ibuprofen in NaOH using MWCNTs/Ag-ZnO/GCE at 30 s deposition time. (B) Plot of I_p vs. deposition potential; Figure S4: (A) Effect of deposition time on the peak current of 0.09 mM ibuprofen using MWNTs/Ag-ZnO/GCE. (B) Plot between Ip vs. accumulation time; Table S3: The values of current for total drug (I_{TD}) free drug (I_{FD}) and the difference of the two currents ΔI; Table S4: Data employed for the determination of function Φ and for the construction of plot Φ versus log [FD] in the case of the IBP-BSA complex in Figure 10.

Author Contributions: M.D. performed the experiments and wrote the manuscript. S.M. contributed to the interpretation of results and manuscript revision. A.S. supervised the research project. All authors have read and agreed to the published version of the manuscript.

Funding: This research received no external funding.

Institutional Review Board Statement: Not applicable.

Informed Consent Statement: Not applicable.

Data Availability Statement: The data presented in this study are available within the article and supplementary material.

Acknowledgments: The authors gratefully acknowledge the laboratory facilities provided by Quaid-i-Azam University, Islamabad, Pakistan to carry out the experimental work.

Conflicts of Interest: The authors declare no conflict of interest.

Sample Availability: Samples of the compounds are available from the authors.

References

1. Pereira, J.C. Environmental issues and international relations, a new global (dis) order-the role of international relations in promoting a concerted international system. *Rev. Bras. Politica Int.* **2015**, *58*, 191–209. [CrossRef]
2. Velusamy, S.; Roy, A.; Sundaram, S.; Mallick, T.K. A review on heavy metal ions and containing dyes removal through graphene oxide-based adsorption strategies for textile wastewater treatment. *J.Chem. Rec.* **2021**, *21*, 1570–1610. [CrossRef] [PubMed]
3. Anju, A.; Ravi, S.P.; Bechan, S. Water pollution with special reference to pesticide contamination in India. *J. Water Resour. Prot.* **2010**, *2*, 17.
4. Grant, R.; Combs, A.; Acosta, D. Experimental models for the investigation of toxicological mechanisms. *Elsevier Sci.* **2010**, *73*, 203–224.
5. Kress, J.P.; Gehlbach, B.; Lacy, M.; Pliskin, N.; Pohlman, A.S.; Hall, J.B. The long-term psychological effects of daily sedative interruption on critically ill patients. *Am. J. Respir. Crit. Care Med.* **2003**, *168*, 1457–1461. [CrossRef] [PubMed]
6. Taschereau-Dumouchel, V.; Michel, M.; Lau, H.; Hofmann, S.G.; LeDoux, J.P. Putting the "mental" back in "mental disorders": A perspective from research on fear and anxiety. *Mol. Psychiatry* **2022**, *27*, 1322–1330. [CrossRef]

7. Takakura, Y.; Hashida, M. Macromolecular carrier systems for targeted drug delivery: Pharmacokinetic considerations on biodistribution. *J. Pharm. Res.* **1996**, *13*, 820–831.
8. Marchlewicz, A.; Guzik, U.; Wojcieszynska, D. Over-the-counter monocyclic non-steroidal anti-inflammatory drugs in environment sources, risks, biodegradation. *Wat. Air Soil Poll.* **2015**, *226*, 335. [CrossRef]
9. Fent, K.; Weston, A.A.; Caminada, D. Ecotoxicology of human pharmaceuticals. *Aquat. Toxicol. Aquat. Toxicol.* **2006**, *76*, 122–159. [CrossRef]
10. Abbas, A.; Ali, M.; Yosef, A.; Abdalmageed, O.; Shaaban, O.F. Sterility, Can the response to three months ibuprofen in controlling heavy menstrual bleeding with copper intrauterine device be predicted at baseline visit. *Clin. Infect. Dis.* **2017**, *108*, 123–124.
11. Bierma-Zeinstra, S.; Brew, J.; Stoner, K.; Wilson, R.; Kilbourn, A.; Conaghan, P.O. Cartilage, A new lipid formulation of low dose ibuprofen shows non-inferiority to high dose standard ibuprofen the FLARE study (flaring arthralgia relief evaluation in episodic flaring knee pain)–a randomised double-blind study. *Osteoarthr. Cartil.* **2017**, *25*, 1942–1951. [CrossRef] [PubMed]
12. Ju, Z.; Li, M.; Xu, J.; Howell, D.C.; Li, Z.; Chen, F.E. Recent development on COX-2 inhibitors as promising anti-inflammatory agents: The past 10 years. *Acta Pharm. Sin. B* **2022**, *12*, 2790–2807. [CrossRef] [PubMed]
13. Song, Y.; Chai, T.; Yin, Z.; Zhang, X.; Zhang, W.; Qian, Y.; Qiu, J.P. Stereoselective effects of ibuprofen in adult zebrafish (Danio rerio) using UPLC-TOF/MS-based metabolomics. *Environ. Pollut.* **2018**, *241*, 730–739. [CrossRef] [PubMed]
14. Bouissou-Schurtz, C.; Houeto, P.; Guerbet, M.; Bachelot, M.; Casellas, C.; Mauclaire, A.-C.; Panetier, P.; Delval, C.; Masset, D.J.R.T. Pharmacology, Ecological risk assessment of the presence of pharmaceutical residues in a French national water survey. *Regul. Toxicol. Pharmacol.* **2014**, *69*, 296–303. [CrossRef]
15. Kappus, H. Irreversible protein binding of 14C-imipramine in rats in vivo. *Arch. Toxicol.* **1976**, *37*, 75–80. [CrossRef]
16. Nakagawa, H.; Yamamoto, O.; Oikawa, S.; Higuchi, H.; Watanabe, A.; Katoh, N.S. Detection of serum haptoglobin by enzyme-linked immunosorbent assay in cows with fatty liver. *Res. Vet. Sci.* **1997**, *62*, 137–141. [CrossRef]
17. Ravelli, D.; Isernia, P.; Acquarulo, A.; Profumo, A.; Merli, D.C. Voltammetric Determination of Binding Constant and Stoichiometry of Albumin (Human, Bovine, Ovine)–Drug Complexes. *J. Anal. Chem.* **2019**, *91*, 10110–10115. [CrossRef]
18. Ghalkhani, M.; Kaya, S.I.; Bakirhan, N.K.; Ozkan, Y.; Ozkan, S.A. Application of nanomaterials in development of electrochemical sensors and drug delivery systems for anticancer drugs and cancer biomarkers. *Crit. Rev. Anal. Chem.* **2022**, *52*, 481–503. [CrossRef]
19. Ahmad, K.; Shah, A.H.; Adhikari, B.; Rana, U.A.; Vijayaratnam, C.; Muhammad, N.; Shujah, S.; Rauf, A.; Hussain, H.; Badshah, A. pH-dependent redox mechanism and evaluation of kinetic and thermodynamic parameters of a novel anthraquinone. *RSC Adv.* **2014**, *4*, 31657–31665. [CrossRef]
20. Shah, A.; Ullah, A.; Rauf, A.; Rehman, Z.U.; Shujah, S.; Shah, S.M.; Waseem, A. Detailed electrochemical probing of a biologically active isoquinoline. *J. Electrochem. Soc.* **2013**, *160*, 597. [CrossRef]
21. Hung, V.W.; Veloso, A.J.; Chow, A.M.; Ganesh, H.V.; Seo, K.; Kenduezler, E.; Brown, I.R.; Kerman, K. Electrochemical impedance spectroscopy for monitoring caspase-3 activity. *Electrochem. Acta* **2015**, *162*, 79–85. [CrossRef]
22. Mamuru, S.A.; Saki, N.; Bello, D.M.; Dalen, M.B. Square Wave Voltammetric Detection of Nitrite on Platinum Electrode Modified with Moringa oleifera Mediated Biosynthesized Nickel Nanoparticles. *J. Adv. Electrochem.* **2018**, *4*, 168–171. [CrossRef]
23. Amin, M.A.; Abd El-Rehim, S.S.; El-Sherbini, E.; Bayoumi, R. The inhibition of low carbon steel corrosion in hydrochloric acid solutions by succinic acid: Part I. Weight loss, polarization, EIS, PZC, EDX and SEM studies. *Electrochem. Acta* **2007**, *52*, 3588–3600. [CrossRef]
24. Randviir, E.P.; Banks, C.E. Electrochemical impedance spectroscopy: An overview of bioanalytical applications. *Anal. Methods* **2013**, *5*, 1098–1115. [CrossRef]
25. Kokab, T.; Shah, A.; Iftikhar, F.J.; Nisar, J.; Akhter, M.S.; Khan, S.B. Amino acid-fabricated glassy carbon electrode for efficient simultaneous sensing of zinc (II), cadmium (II), copper (II), and mercury (II) ions. *ACS Omega* **2019**, *4*, 22057–22068. [CrossRef]
26. Wang, J.; Yang, B.; Wang, H.; Yang, P.; Du, Y. Highly sensitive electrochemical determination of Sunset Yellow based on gold nanoparticles/graphene electrode. *Anal. Chim. Acta* **2015**, *893*, 41–48. [CrossRef]
27. Akbari, M.; Mohammadnia, M.S.; Ghalkhani, M.; Aghaei, M.; Sohouli, E.; Rahimi-Nasrabadi, M.; Arbabi, M.; Banafshe, H.R.; Sobhani-Nasab, A. Development of an electrochemical fentanyl nanosensor based on MWCNT-HA/Cu-H3BTC nanocomposite. *J. Ind. Eng. Chem.* **2022**, *114*, 418–426. [CrossRef]
28. Suresh, E.; Sundaram, K.; Kavitha, B.; Kumar, S. Square wave voltammetry sensing of ibuprofen on glassy carbon electrode. *Int. J. Pharmtech Res.* **2016**, *9*, 182–188.
29. Montes, R.H.; Lima, A.P.; Cunha, R.R.; Guedes, T.J.; dos Santos, W.T.; Nossol, E.; Richter, E.M.; Munoz, R.A.A. Size effects of multi-walled carbon nanotubes on the electrochemical oxidation of propionic acid derivative drugs: Ibuprofen and naproxen. *J. Electroanal. Chem.* **2016**, *775*, 342–349. [CrossRef]
30. Abbas Momtazi, A.; Sahebkar, A. Difluorinated curcumin: A promising curcumin analogue with improved anti-tumor activity and pharmacokinetic profile. *Curr. Pharm. Des.* **2016**, *22*, 4386–4397. [CrossRef]
31. Mekassa, B.; Tessema, M.; Chandravanshi, B.S.; Tefera, M. Square wave voltammetric determination of ibuprofen at poly (l-aspartic acid) modified glassy carbon electrode. *IEEE Sens. J.* **2017**, *18*, 37–44. [CrossRef]
32. Rivera-Hernandez, S.I.; Alvarez-Romero, G.A.; Corona-Avendano, S.; Páez-Herrnndez, M.E.; Galán-Vidal, C.A.; Romero-Romo, M. Technology, Voltammetric determination of ibuprofen using a carbon paste–multiwalled carbon nanotube composite electrode. *Instru. Sci. Technol.* **2016**, *44*, 483–494. [CrossRef]

33. Ţuchiu, B.-M.; Stefan-van Staden, R.-I.; van Staden, J. Recent Trends in Ibuprofen and Ketoprofen Electrochemical Quantification–A Review. *Crit. Rev. Anal. Chem.* **2022**, 1–12. [CrossRef] [PubMed]
34. Motoc, S.; Manea, F.; Iacob, A.; Martinez-Joaristi, A.; Gascon, J.; Pop, A.; Schoonman, J.J.S. Electrochemical selective and simultaneous detection of diclofenac and ibuprofen in aqueous solution using HKUST-1 metal-organic framework-carbon nanofiber composite electrode. *Sensors* **2016**, *16*, 1719. [CrossRef] [PubMed]
35. Svorc, L'.; Strezova, I.; Kianickova, K.; Stankovic, D.M.; Otrisal, P.; Samphao, A. An advanced approach for electrochemical sensing of ibuprofen in pharmaceuticals and human urine samples using a bare boron-doped diamond electrode. *J. Electroanal. Chem.* **2018**, *822*, 144–152. [CrossRef]
36. Nair, A.S.; Sooraj, M. Molecular imprinted polymer-wrapped AgNPs-decorated acid-functionalized graphene oxide as a potent electrochemical sensor for ibuprofen. *J. Mater. Sci.* **2020**, *55*, 3700–3711. [CrossRef]
37. Vuignier, K.; Schappler, J.; Veuthey, J.L.; Carrupt, P.-A.; Martel, S. Drug–protein binding: A critical review of analytical tools. *Anal. Bioanal. Chem.* **2010**, *398*, 53–66. [CrossRef] [PubMed]
38. Leuna, J.B.M.; Sop, S.K.; Makota, S.; Njanja, E.; Ebelle, T.C.; Azebaze, A.G.; Ngameni, E.; Nassi, A. Voltammetric behavior of Mammeisin (MA) at a glassy carbon electrode and its interaction with Bovine Serum Albumin (BSA). *Bioelectrochemistry* **2018**, *119*, 20–25. [CrossRef] [PubMed]
39. Gupta, V.K.; Jain, R.; Radhapyari, K.; Jadon, N.; Agarwal, S.B. Voltammetric techniques for the assay of pharmaceuticals—A review. *Anal. Biochem.* **2011**, *408*, 179. [CrossRef]
40. Maruthamuthu, M.; Kishore, S. Binding of ketoprofen with bovine serum albumin. *Proc. Indian Acad. Sci.-Chem. Sci.* **1987**, *99*, 187–193. [CrossRef]
41. Pacifici, G.; Viani, A.; Schulz, H.U.; Frercks, H.P. Plasma protein binding of furosemide in the elderly. *Eur. J. Clin. Pharmacol.* **1987**, *32*, 199–202. [CrossRef] [PubMed]
42. Prasanth, S.; Sudarsanakumar, C. Elucidating the interaction of L-cysteine-capped selenium nanoparticles and human serum albumin: Spectroscopic and thermodynamic analysis. *New J. Chem.* **2017**, *41*, 9521–9530. [CrossRef]
43. Wang, X.; Liu, Y.; He, L.L.; Liu, B.; Zhang, S.Y.; Ye, X.; Jing, J.J.; Zhang, J.-F.; Gao, M.J.F.; Toxicology, C. Spectroscopic investigation on the food components–drug interaction: The influence of flavonoids on the affinity of nifedipine to human serum albumin. *Food Chem. Toxicol.* **2015**, *78*, 42–51. [CrossRef] [PubMed]
44. Kariv, I.; Cao, H.; Oldenburg, K.R. Development of a high throughput equilibrium dialysis method. *J. Pharm. Sci.* **2001**, *90*, 580–587. [CrossRef]

Disclaimer/Publisher's Note: The statements, opinions and data contained in all publications are solely those of the individual author(s) and contributor(s) and not of MDPI and/or the editor(s). MDPI and/or the editor(s) disclaim responsibility for any injury to people or property resulting from any ideas, methods, instructions or products referred to in the content.

Article

Chemico-Physical Properties of Some 1,1′-Bis-alkyl-2,2′-hexane-1,6-diyl-bispyridinium Chlorides Hydrogenated and Partially Fluorinated for Gene Delivery

Michele Massa [1], Mirko Rivara [2], Thelma A. Pertinhez [3], Carlotta Compari [2], Gaetano Donofrio [4], Luigi Cristofolini [5], Davide Orsi [5], Valentina Franceschi [4] and Emilia Fisicaro [2,*]

1. Department of Maternal Infantile and Urological Sciences, Sapienza University of Rome, 00165 Rome, Italy; michele.massa@uniroma1.it
2. Department Food and Drug, University of Parma, Parco Area delle Scienze, 27/A, 43124 Parma, Italy; mirko.rivara@unipr.it (M.R.); carlotta.compari@unipr.it (C.C.)
3. Department of Medicine and Surgery, University of Parma, Via Volturno, 39, 43125 Parma, Italy; thelma.deaguiarpertinhez@unipr.it
4. Department of Veterinary Sciences, University of Parma, Via del Taglio, 10, 43126 Parma, Italy; gaetano.donofrio@unipr.it (G.D.); valentina.franceschi@unipr.it (V.F.)
5. Department of Mathematical, Physical and Computer Sciences, University of Parma, Parco Area delle Scienze 7/a, 43124 Parma, Italy; luigi.cristofolini@unipr.it (L.C.); davide.orsi@unipr.it (D.O.)
* Correspondence: emilia.fisicaro@unipr.it

Abstract: The development of very efficient and safe non-viral vectors, constituted mainly by cationic lipids bearing multiple charges, is a landmark for in vivo gene-based medicine. To understand the effect of the hydrophobic chain's length, we here report the synthesis, and the chemico-physical and biological characterization, of a new term of the homologous series of hydrogenated *gemini* bispyridinium surfactants, the 1,1′-bis-dodecyl-2,2′-hexane-1,6-diyl-bispyridinium chloride (GP12_6). Moreover, we have collected and compared the thermodynamic micellization parameters (cmc, changes in enthalpy, free energy, and entropy of micellization) obtained by isothermal titration calorimetry (ITC) experiments for hydrogenated surfactants GP12_6 and GP16_6, and for the partially fluorinated ones, FGPn (where n is the spacer length). The data obtained for GP12_6 by EMSA, MTT, transient transfection assays, and AFM imaging show that in this class of compounds, the gene delivery ability strictly depends on the spacer length but barely on the hydrophobic tail length. CD spectra have been shown to be a useful tool to verify the formation of lipoplexes due to the presence of a "tail" in the 288–320 nm region attributed to a chiroptical feature named ψ-phase. Ellipsometric measurements suggest that FGP6 and FGP8 (showing a very interesting gene delivery activity, when formulated with DOPE) act in a very similar way, and dissimilar from FGP4, exactly as in the case of transfection, and confirm the hypothesis suggested by previously obtained thermodynamic data about the requirement of a proper length of the spacer to allow the molecule to form a sort of molecular tong able to intercalate DNA.

Keywords: heterocyclic *gemini* cationic surfactants; non-viral vectors; gene delivery; partially fluorinated *gemini* surfactants; atomic force microscopy on DNA; DNA-surfactant interaction; DNA circular dichroism spectra; ellipsometry

1. Introduction

The relevant practical interest of *gemini* surfactants, i.e., surfactants consisting of at least two identical hydrophobic chains and two polar head groups covalently bound together by a spacer, is due to their enhanced surface properties in respect to the monomeric counterparts [1–5]. Taking advantage of these properties, their application in biomedical and pharmaceutical fields as new drug delivery systems and as non-viral vectors for gene delivery has sparked great interest. The internalization of genetic material inside mammalian cells is only possible using vectors, both viral and non-viral. Non-viral vectors are constituted by cationic lipids able to bind and compact the genetic material into soft nanoparticles of tuneable size, so that it is protected against the action of endo- and exonucleases present in physiological fluids. Very recently, the spreading worldwide of the SARS-CoV-2 virus has focused scientific and economic efforts on searching for vaccines to fight the virus. DNA- and RNA-based vaccines, able to express the spike protein when internalized in cells, are now the most-used around the world [6,7]. Non-viral vectors are in many cases preferred, particularly when RNA-based, which require only to overcome the external cell membrane, instead of DNA-based which, needing to reach the nucleus for expression, must be introduced inside the cells. In fact, viral vectors, notwithstanding their superior efficiency, could give rise to adverse or immunogenic reactions or replications, limiting their use. Biomedical and biopharmaceutical applications of cationic *gemini* amphiphiles as gene delivery vectors, together with their chemico-physical and aggregation properties and the effect on the efficiency of transfection of their chemical structure have been extensively reviewed [5,8,9]. Non-viral vectors, obtained by synthetic route, have the advantage of being more reproducible than viral ones, and give us the possibility to understand the effect of the various moieties constituting the molecule and, eventually, to add to the vector a chemical group that can be recognized by the target cell. For many years our research was devoted from a synthetic, thermodynamic, and biomedical point of view to the study of new cationic *gemini* surfactants, having as polar head two pyridinium moieties, with the aim to find out structure–activity relationships useful for the optimization of their gene delivery ability, through the modulation of the lengths of the hydrophobic chains and spacer, counterion, and other structural modifications [10–18]. Taking advantage of this opportunity, we have considered not only hydrogenated *gemini* surfactants, but also the corresponding partially fluorinated (otherwise called "hybrid surfactants") with the idea of obtaining non-viral vectors able to protect the genetic material also in those biological fluids containing endogenous hydrogenated interfering surfactants, as pulmonary surfactants or bile salts, able to destroy the lipoplexes before they enter inside the diseased cells [19–21]. Fluorinated surfactants are at the same time highly hydrophobic and lipophobic due to the structure of the fluorine atoms having larger van der Waals radii and lower polarizability than the hydrogen atoms, and do not form mixed micelles with hydrogenated surfactants [22–24]. Fluorination of cationic lipids has been proposed, among others, in the treatment of cystic fibrosis and cystic fibrosis-associated diseases [20]. Initially, we have studied the solution thermodynamics of new compounds having spacers of three, four, eight, and twelve carbon atoms, using direct methods and paying attention to the trends of partial molar volumes, compressibilities, and enthalpies vs. concentration [10–18]. Particularly, the measurement of their solution enthalpies gives us the key for understanding their transfection activity. Between the hydrogenated compounds, those with spacers formed by four carbon atoms shows unexpected enthalpic properties vs. concentration, which was explained by a conformation change of the molecule in solution due to stacking interactions between the two pyridinium rings. In this way, the molecule behaves as a molecular tong, able to grip the DNA bases, giving rise to a transfection activity comparable to that of the commercial reagent. If the hydrophobic chain is modified by partial fluorination, a greater length of the spacer is needed to fold the molecule. The comparison with the hydrogenated analogues reveals a greater ability of the partially fluorinated compounds to compact DNA, but only in presence of DOPE. In a recent paper [18], we tried to understand if the transfection ability is limited to a given value of spacer length, or spans over a range of values, by synthesizing the compounds, both hydrogenated and partially fluorinated, with spacer constituted by six methylene groups, namely, 1,1′-bis-hexadecyl-2,2′-hexane-1,6-diyl-bis (pyridinium) chloride (GP16_6) and 1,1′-bis(3,3,4,4,5,5,6,6,7,7,8,8,8-tridecafluorooctyl)-2,2′-hexane-1,6-diyl-bis (pyridinium) chloride (FGP6) to fill the gap between active and non-active compounds. We have outlined the completely different behaviour between hydrogenated and fluorinated compounds:

in the case of hydrogenated compounds, only spacer four can deliver genes inside the cell, independently of the use of DOPE. On the contrary, fluorinated compounds having a spacer with six or eight methylene groups give rise to a very high transfection activity only in the presence of DOPE. This result indicates that the spacer length appropriate for activity falls in a broader, but always limited, range of values (spacer 12 is inactive for both classes of compounds). It is known that the critical micelle concentrations (cmc) of the fluorinated surfactants, determined by the balance of the hydrophobicity and hydrophilicity of the molecule, are close to those of ordinary surfactants whose hydrocarbon chain lengths are about 1.5 times longer [3]. This means that the -CF_2 group is 1.5 times more hydrophobic than the -CH_2 group. To compare in a correct way the behaviour of FGP6 with the corresponding hydrogenated surfactants having a similar hydrophobicity of the alkyl chain, we synthesized the new compound 1,1'-bis-dodecyl-2,2'-hexane-1,6-diyl-bis (pyridinium) chloride, (GP12_6), similar to GP16_6, but with an alkyl chain 12 carbon atoms long. The present paper reports the chemico-physical and tensidic properties of the compounds with spacer 6, together with the synthesis and the biological characterization of the new compound 1,1'-bis-dodecyl-2,2'-hexane-1,6-diyl-bis (pyridinium) chloride (GP12_6), to evaluate the effect of the hydrophobic chain length.

2. Results and Discussion

2.1. Micelle Formation Thermodynamics

Isothermal titration calorimetry (ITC) constitutes a particularly useful technique for studying thermodynamics of micelle formation. In fact, from the same titration experiment the thermodynamic parameters of micelle formation can be extracted, namely the critical micelle concentration (cmc), in relation to the change in free energy of micellization (ΔG_{mic}), and the enthalpy of micelle formation (ΔH_{mic}). The change in micellization entropy (ΔS_{mic}) is obtained by difference. By adding the surfactant solution at a concentration of least ten time the cmc in the calorimetric cell containing water, micelles are destroyed, and the heat observed is mainly relative to the process of micelle disruption. Above the cmc region, the heat due to the process of micelle dilution is registered. The micellization parameters are obtained by extrapolating at the cmc the trends of dilution enthalpies before and after cmc, i.e., by applying a pseudo-phase transition model, in which the aggregation process is considered as a phase transition, taking place at equilibrium. In the literature [25], two different ways of obtaining the cmc values from ITC measurements are reported: (1) as the concentration of the crossing point between extrapolated initial and linear ascent lines where 1% of the surfactants is in micellar form; (2) as the inflection point in the titration curve. Both methods give about the same value when the surfactant micelles have a great aggregation number, but it is not so when the transition between monomers and micelle zone is not sharp. We have chosen, in agreement with the greatest part of the data reported in the literature, to find out the cmc as the inflection point of the titration curve, notwithstanding the not-straightforward detection. In our previous studies of solution thermodynamics of this class of *gemini* surfactants, we have measured the dilution enthalpies using a batch flow calorimeter diluting 1:1 with water the surfactant solutions prepared at different concentrations, and expressing the experimental data in terms of apparent and partial molar quantities of the solute referred to the infinite dilution as reference state [12,13,26]. When the curves of the dilution enthalpies vs. concentration obtained from calorimetric titration experiments are directly used to extract the thermodynamic micellization parameters, the reference state is the concentration of the titrating solution; this explains why, using different starting concentrations, different micellization enthalpies are obtained [27].

Therefore, ITC experiments, both for partially fluorinated and hydrogenated compounds, were performed by diluting a solution of 20 mM of surfactants so that the comparisons between the compounds here studied are meaningful, being referred to the same standard state. Figure 1 shows the ITC output for the compounds under investigation and the corresponding plots of the dilution enthalpies vs. the surfactant concentration in the calorimetric cell, starting from a titrating solution 20 mM, i.e., the trends of apparent molar enthalpies with reference state of 20 mM. From these curves, the cmc and the micelle formation enthalpy, ΔH_{mic}, are obtained using a pseudo-phase transition model. The micellization free energy, ΔG_{mic}, is calculated from equation:

$$\Delta G_{\text{mic}} = RT \ln \text{cmc}$$

where the cmc is expressed as mole fraction, without considering the degree of counterion binding, β [28]. Moreover,

$$\Delta S_{\text{mic}} = (\Delta G_{\text{mic}} - \Delta H_{\text{mic}})/T$$

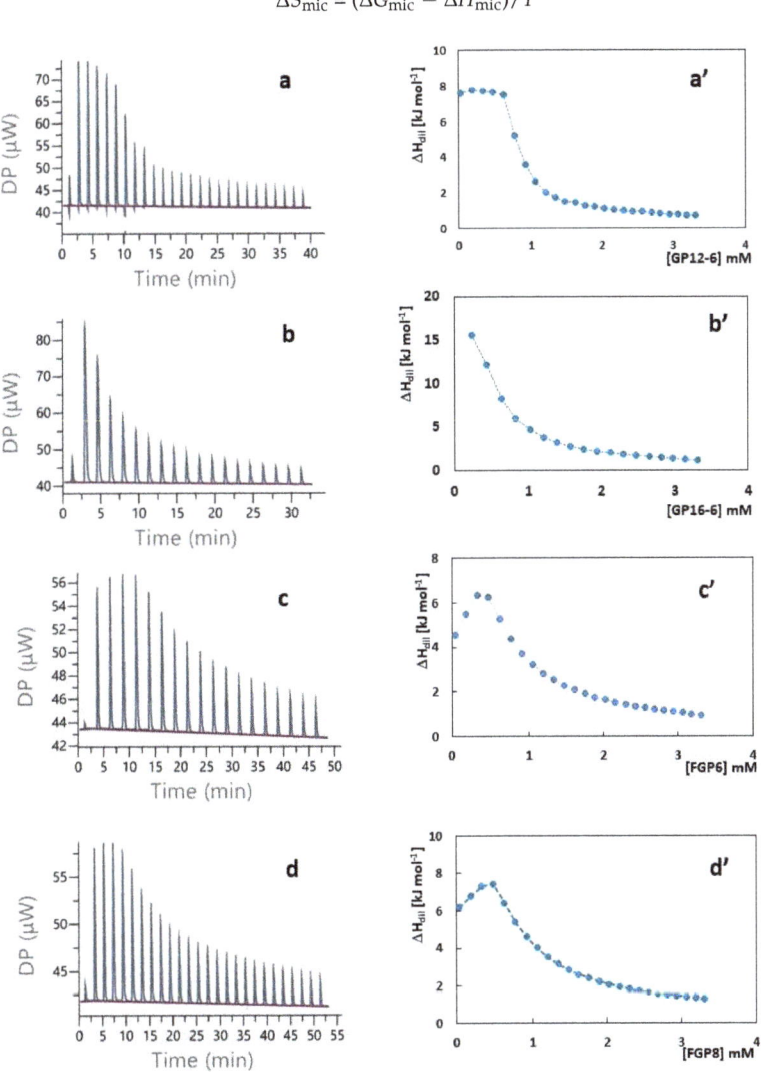

Figure 1. Heat rate (μW) vs. time profiles obtained by injecting a 20 mM surfactant solution into a 200 μL reaction cell filled with water and (specified by a single quote mark) the dependence of the dilution enthalpy change vs. surfactant concentration in the reaction cell for: (**a**) GP12_6; (**b**) GP16_6; (**c**) FGP6; (**d**) FGP8.

The cmc values and the thermodynamic parameters for the micellization process so obtained are reported in Table 1.

Table 1. Thermodynamic parameters for micelle formation of the surfactants under study.

	cmc [mM]	$x_{cmc}\, 10^5$	ΔH_{mic} [kJ mol^{-1}]	ΔG_{mic} [kJ mol^{-1}]	ΔS_{mic} [J K^{-1} mol^{-1}]
FGP6	0.89	1.603	−6.1	−27.4	71
FGP8	0.77	1.387	−6.2	−27.7	72
GP12-6	0.92	1.657	−6.5	−27.3	70
GP16-6	0.41	0.738	−14.0	−29.3	51

Data shown in Table 1 confirm the similarity in hydrophobicity of FPG6 and GP12_6, allowing us to attribute the different behaviour in gene delivery ability exclusively to the fluorinated moieties. Moreover, we recall that thermodynamic properties of hydrogenated and fluorinated surfactants in solution, strictly related to the ability to compact DNA, depend both on the difference in size between fluorine and hydrogen atoms (van der Waals radii, F = 1.35 and H = 1.2 Å) and on the difference in electronegativity (Pauling scale, F = 4.0 and H = 2.1) [24].

2.2. Biological Assays of GP12_6

As outlined before, we have found that the partial substitution of the hydrogenated chains by shorter fluorinated ones gives rise to a very interesting gene delivery ability, as for FGP6 [18] and FGP8 [17]. Because it is known that the -CF$_2$ group is 1.5 times more hydrophobic than the -CH$_2$ group [29], confirmed by the data reported in Table 1, for a correct comparison of the behaviour of FGP6 with the corresponding hydrogenated surfactant, it makes sense to report here the biological data relative to the newly synthesized GP12_6.

As already conducted for the previously synthesized terms of this series [17,18], we started by testing the cytotoxicity of GP12_6 on RD4 and on A549 cells by an MTT proliferation assay. Results are shown in Figure 2 in comparison with GP16_6 [18].

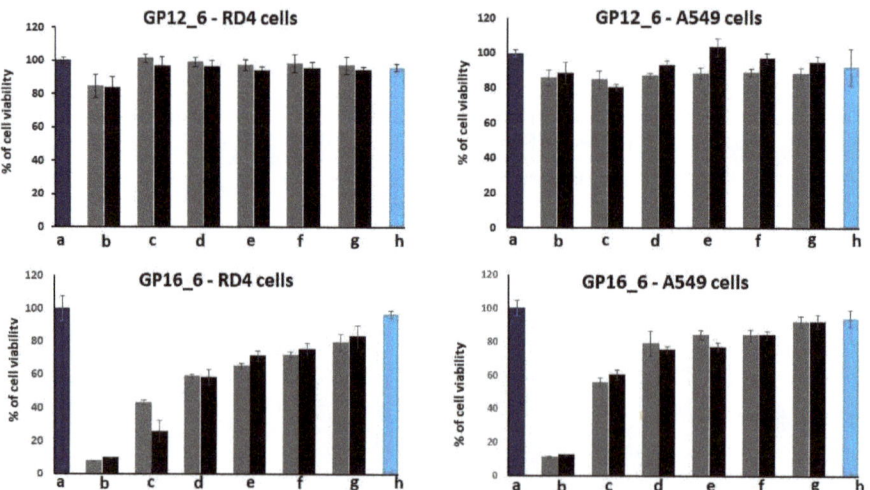

Figure 2. MTT tests on RD4 (left) and A549 cells (right) for GP12_6 (upper) and GP16_6 (lower) from ref. [18]: (**a**) untreated cells; (**b**) 40 µM; (**c**) 20 µM; (**d**) 10 µM; (**e**) 5 µM; (**f**) 2.50 µM; (**g**) 1.25 µM; (**h**) only DOPE. In gray, surfactant alone; in black, surfactant:DOPE = 1:2. Values are the mean ± S.D. of three independent experiments (n = 8 per treatment, p < 0.05).

By analogy with similar compounds previously studied [18], we have tested the newly synthesized compound on the human rhabdomyosarcoma RD4 cell line and on adenocarcinoma human alveolar basal epithelial cells A549, used as models for the study of lung cancer and for the development of drug therapies against it. Because this information is needed for planning transfection experiments, we performed our cytotoxicity assays also in presence of DOPE, which is essential in transfection experiments. Results in Figure 2 clearly show that the cytotoxicity of GP12_6 is very low on both kinds of cell lines, without significant differences with and without DOPE and that the cytotoxicity increases with increasing the hydrophobic chain length. The surfactant with longer tails interacts more strongly with the cellular membrane, and their complexes with the DNA could penetrate inside the cell more easily. However, their action is hindered by the increased cytotoxicity for gene delivery purposes. In fact, it is reported that the biological activity of cationic surfactants increases with the chain length up to a critical point [30]. In the case of the homologous series of alkanediyl-α,ω-bis-(dimethylalkylammonium bromide), the member with 2 alkyl chains of 16 carbon atoms is generally the most biologically active [31].

Information about the ability of the compounds under investigation to interact with DNA is obtained with EMSA experiments. Figure 3 upper shows that GP12_6 does not shift the DNA, not even at the highest concentration used, whereas GP16_ shifts the DNA at a concentration of 200 µM.

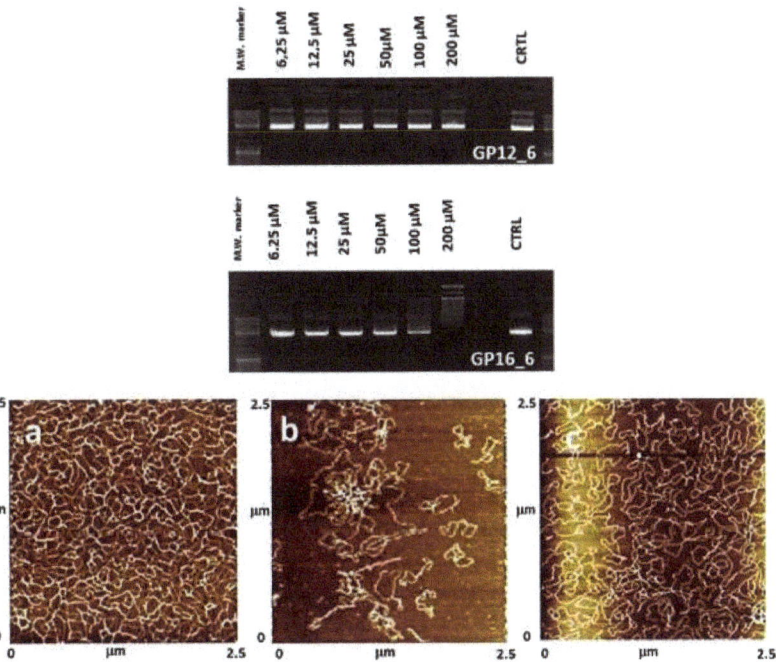

Figure 3. Upper: EMSA experiments showing complexation of GP12_6 with circular plasmid pEGFP-C1 as a function of surfactant concentration (µM), compared with GP16_6 [18]. Only the plasmid was used as a negative control, which is completely unshifted. Cationic lipid 100 µM corresponds to N/P ≈ 1. Lower: AFM images of DNA plasmid incubated with the hydrogenated compounds: (**a**) plasmid + GP12_6; (**b**) plasmid + GP16_6; (**c**) plasmid alone as control. All images were obtained with supercoiled 0.5 nM pEGFP-C1 plasmid deposited onto mica and with the microscope operating in tapping mode in the air at N/P ratio = 0.5.

EMSA results are confirmed by morphological study by AFM in tapping mode (Figure 3 lower), using circular DNA, as described in the experimental section. The plasmid DNA alone deposited onto freshly cleaved mica was imaged as control. The images in the presence of the cationic surfactants were taken at a N/P ratio = 0.5, corresponding to the highest concentration of surfactant used in EMSA experiments. As expected, AFM images confirm that the hydrogenated surfactants with a spacer six methylene long cannot compact enough DNA in nanoparticles. In the presence of GP12_6, the AFM image is unmodified in respect to DNA alone, whereas GP16_6 starts to give rise to very loose aggregates.

The ability of GP12_6 to deliver DNA inside cells was studied with a transient transfection assay. Results on RD4 and A549 cells as a function of concentration are reported in Supplementary Materials (Figures S5 and S6). The experiments were conducted with the surfactant alone and with surfactant:DOPE = 1:2, because previous experiments [17,18] have shown how the presence of DOPE can greatly enhance transfection when fluorinated surfactants are used to form lipoplexes. According to EMSA and AFM data, GP12_6 cannot act as non-viral vector both in presence and in the absence of DOPE, as GP16_6.

2.3. CD Spectroscopy

For better understanding the ability of the compounds under study to interact with DNA, we have undergone CD measurements, which are very sensitive to structural changes in biological molecules.

The pEGFP-C1 plasmid DNA far UV spectrum (220–320 nm) shows the contribution of right-handed A- and B-forms: where the positive band is due to base stacking, and the negative band is due to the polynucleotide right-handed helicity [32] (Figure 4). The A-form is characterized by an intense positive band around 270 nm and a weak negative band near 235 nm, and the B-form by a positive band around 280 nm and a negative at 240 nm.

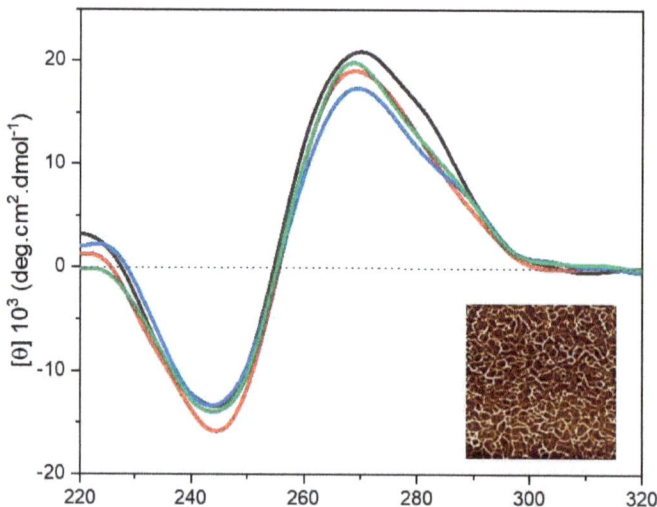

Figure 4. CD spectra of DNA (black), and in the presence of GP12_6 (N/P = 1 red, N/P = 0.5 blue, N/P = 0.25 green). In the insert: AFM image for N/P = 0.5 (see also Figure 3).

Changes in the polynucleotide CD secondary structure upon surfactant addition result from their interactions. We kept the concentration of pEGFP-C1 plasmid DNA constant, and the effect of variable amounts of the 6-spacer surfactants GP16-6, GP12-6 and FGP8-6 (molar ratio 1:1, 1:2 and 1:4) were evaluated.

Upon the addition of surfactant to the DNA solution, different spectra were observed depending on the surfactant type, the length of the chain, hydrophobicity, and the molar ratio.

A weak interaction was observed adding GP12-6 at all molar ratios, which produced a shift and decrease in the positive band at 280 nm, attributed to the B-form (Figure 4).

GP16-6 affects the polynucleotide base stacking primarily. The spectra (Figure 5) show the resolution of the convoluted original positive band of the two positive bands characteristic of the A- (270 nm) and B- (283 nm) forms, respectively. We interpret this change in the CD spectrum as the result of a decrease in the A-form and an increase in the B-form.

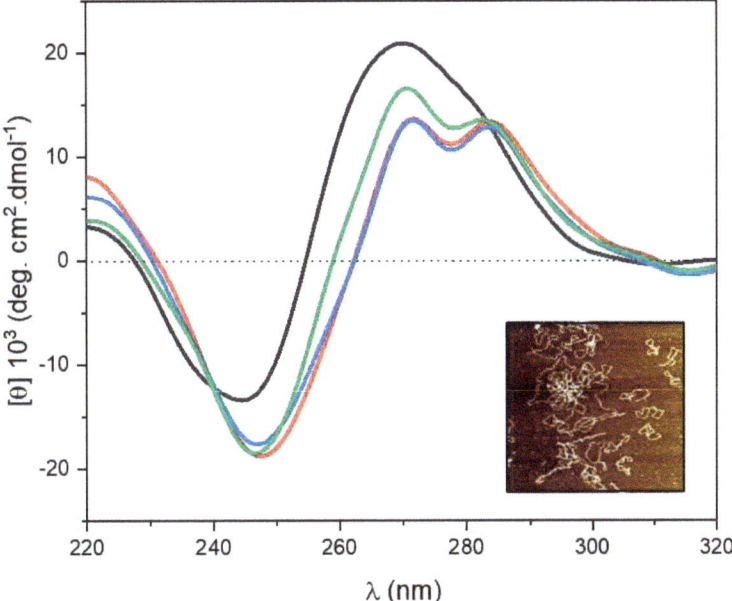

Figure 5. CD spectra of DNA (black), and in the presence of GP16_6 (N/P = 1 red, N/P = 0.5 blue, N/P = 0.25 green). In the insert: AFM image for N/P = 0.5 (see also Figure 3).

Differently, the interaction with FGP8-6 is more complex and depends on the molar ratio (Figure 6). At a 1:1 molar ratio, the A-form decreases, as suggested by the reduction of its characteristic bands. At a molar ratio of 1:2, there is an evident shift of the negative band to the B-form and the convolution of the A- and B-form gives rise to positive bands. At a higher molar ratio (1:4) a clear increase in the B-form (278 nm) is observed. The appearance of a positive band at 288 nm wavelength (at 1:2 and 1:4 molar ratio) indicates DNA compaction [33]. The "tail" in the 288–320 nm region depends on the presence of FGP6 and it is attributed to a chiroptical feature named ψ-phase. Such a phase is a consequence of the DNA structural transition from the B-form induced by the surfactant and reflects the supramolecular structure of the complex [34]. The plasmid is tightly packed together, forming a highly condensed structure [35,36] as shown in the AFM image reported in ref. [18]. In Supplementary Materials the deconvolutions of CD spectra (Figure S7) and the table (Table S1) with the bands obtained using a gaussian model with Origin-Pro 2021 are reported for clarity.

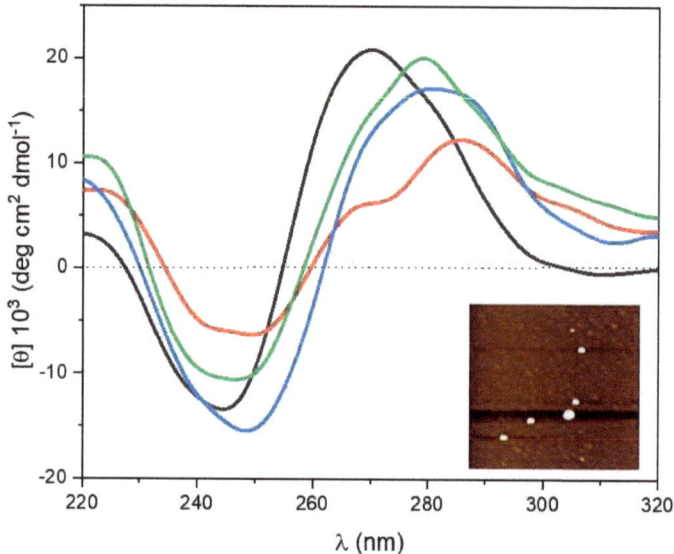

Figure 6. CD spectra of DNA (black), and in the presence of FGP6 (N/P = 1 red, N/P = 0.5 blue, N/P = 0.25 green). In the insert: AFM image for N/P = 0.5 from ref. [18].

2.4. Ellipsometry and Surface Pressure

Because the partially fluorinated surfactants are the more interesting surfactants for gene delivery purposes, we focalized our attention on the study of FGP4 (lacking transfection ability), and FGP6 and FGP8 (with a very noteworthy transfection ability) as a function of the spacer length. Initially, we studied the kinetics of surface layer formation to find out how much time these compounds take to reach equilibrium. Kinetics evolved over several hours (see Supplementary Materials, Figures S8 and S9). FGP4 100 µM reached equilibrium after about 6 h, while equilibrium of FGP4 50 µM was not reached even after 15 h. Due to this slow tendency to reach equilibrium, we decided to make the lowering of surface tension and ellipsometry measurements after two hours, a range of time that is compatible with their persistence in the body, and at the same concentrations used for the transfection and cytotoxicity assays, i.e., 2.5 µM, 5 µM, 10 µM and 20 µM, concentrations well below the cmc. Therefore, all experiments were performed at a partial coverage of the surface due to the low concentrations we are interested in, and the time chosen for the measurement, relevant for the transfection experiment but not enough to reach equilibrium.

Under the above conditions, by ellipsometry at the incidence angle $\Phi = 55°$, we measured the optical thickness of the film that is spontaneously formed at the initially clean air-water interface. We report in Figure 7 the variation of phase angle Δ, with respect to the value of the bare air/water interface Δ_0, as a function of time and of surfactant bulk concentration. At the highest concentrations of surfactant studied, this amounts to roughly the same value for all the molecules, of the order of a variation of $\Delta - \Delta_0 \approx -1.5°$.

The small film thickness ensures we are within the Drude approximation. In this case, the variation of Δ is proportional to the film thickness multiplied by a factor proportional to the difference between the refractive index of the film and that of the subphase ($n = 1.33$ for water as in the present case). It is, therefore, crucial to know the film's refractive index. Unfortunately, there are no data in the literature as these are newly synthesized molecules, and it is difficult to predict it a priori.

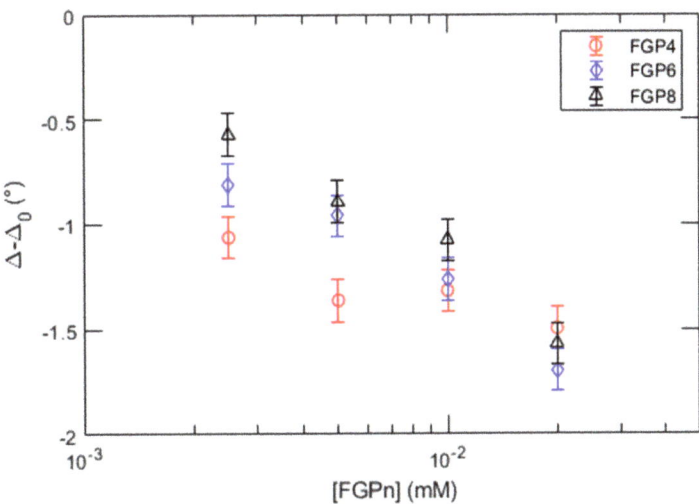

Figure 7. The variation of the ellipsometric phase angle Δ as a function of surfactant concentration, after 2 h of incubation, for the three surfactant molecules studied. Note the semilogarithmic scale.

A first estimate can be drawn based on the molecular structure: FGPs molecules contain two types of units, fluorinated and hydrogenated. Fluorinated alkyl chains, such as perfluorohexanoic acid (CAS 307-24-4) have $n = 1.301$, pyridine has $n = 1.51$, while alkyl chains have typically $n \simeq 1.40$ at the relevant wavelength. We can attempt some rough volumetric considerations, assuming the polar head comprising of pyridine and alkyl spacer with thickness $\sim$ 3–4 Å, and average refractive index 1.46–1.47 depending on alkyl spacer length, while the *gemini* partially fluorinated chains, common to all the molecules, account for $\sim$ 2 + 6 Å of thickness and refractive index $n = 1.40$ and $n = 1.30$ for the protonated and fluorinated chains, respectively.

Basing on such very broad considerations, we can estimate the molecular refractive index to be not too far from $n = 1.37$–1.39. With these values, film thickness results to be of the order of 15–20 Å, thus compatible with the formation of a monolayer at the interface. A more accurate estimate of film thickness could be made by using X-ray reflectometry, while neutron reflectometry could provide even more detailed insights into the film structure [37].

The relative weight of the fluorinated part (small refractive index n) in respect to the hydrogenated part (big refractive index n) is larger in FGP4 than in FGP6 and FGP8. This suggests that the refractive indexes of FGP4 is the smallest and closest to water, followed by FGP6, while FGP8 is the largest. This explains, in part, why FGP4 has the smallest variation of Δ.

At lower concentrations, however, the different molecules exhibit a very different behaviour, with FGP6 and FGP8 displaying a lower variation of Δ than FGP4. This cannot be accounted for by the previous argument on the contribution of fluorinated chains to the refractive index (which would predict a smaller variation for FGP4) and can only be rationalized assuming that the films formed by different molecules have different thicknesses: in films formed by FGP6 and FGP8 at low packing, thanks to the longer alkyl spacer, the fluorinated chains may lay oblique and close to the water surface, while on the contrary in FGP4 even at low coverage the molecules are forced to stand more vertical due to the shorter alkyl spacer.

Coherently with this explanation, we also find that in this regime of lower concentration, the surface pressure Π, i.e., the reduction of interfacial tension from its value for the bare air-water interface $\gamma = 72.6$ mN/m, is greater for FGP6 and FGP8 than for FGP4, especially at the lower concentrations (Figure 8).

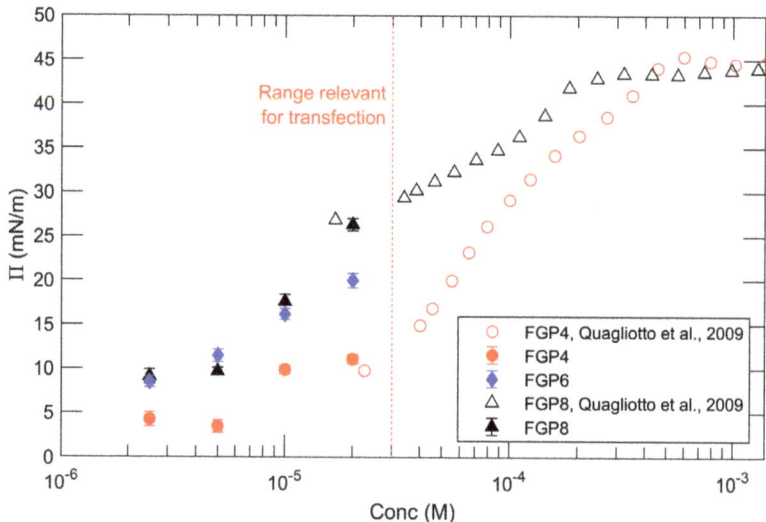

Figure 8. Variation of the surface pressure Π as a function of surfactant concentration in the bulk, after 2 h of incubation, for the 3 surfactant molecules studied (filled symbols); for FGP4 and FGP8, tensiometry data from ref. [16] have been included (empty symbols). Note the semilogarithmic scale. The vertical dashed line indicates the maximum concentration considered for the formation of non-viral gene delivery vectors.

In the same Figure, we also show the limiting pressures that would be reached by the same surfactants at much higher concentrations, above their respective CMC values.

In the same Figure, we also report previously published data for FGP4 and FGP8 at concentration up to and above cmc, adapted from ref. [16]. In summary, it is also important to notice that FGP6 and FGP8 act in a very similar way, and dissimilar from FGP4, exactly as in the case of transfection. This suggests the presence of a dichotomy: either the spacer is or is not long enough for the flexibility required by the molecule to accommodate at the interface in the present case, or around a DNA fragment in the case of interest for transfection.

3. Materials and Methods

3.1. Compounds

All the compounds under study with six methylene spacers, hydrogenated GPm_6, where m il the hydrophobic tail length (m = 12 and 16), and partially fluorinated FGPn (where n is the spacer length, n = 4, 6, 8) with hydrophobic tails constituted by two -$(CH_2)_2$-$(CF_2)_5$-CF_3 groups, were prepared by us. The synthesis of GP16_6 and FGP6 are reported in ref. [18], that of FGP4 and FGP8 in ref. [16]. We synthesized for the first time GP12_6, having a hydrophobicity comparable to FGP6 (see text), and its synthesis is here reported. The structure of GP12_6 and of the compounds with spacer six methylene long under study is shown in Figure 9.

3.2. Chemical Synthesis

We synthesized the new compound 1,1′-bis-dodecyl-2,2′-hexane-1,6-diyl-bispyridinium chloride (GP12_6), following and adapting, when necessary, the procedures previously reported [11,16].

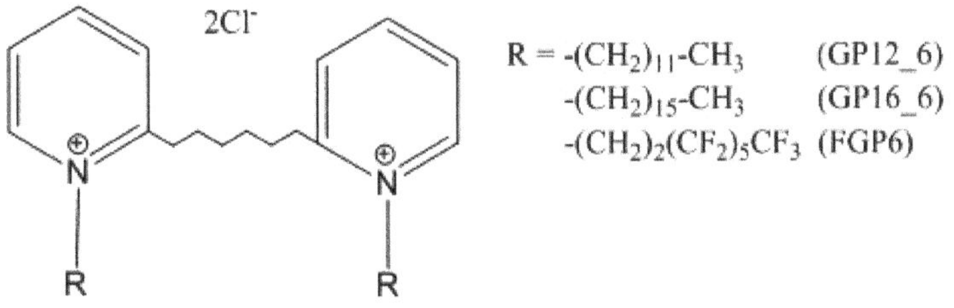

Figure 9. Chemical structure of the compounds under study with spacer six methylene long.

Synthesis of 1,1′-bis-dodecyl-2,2′-hexane-1,6-diyl-bispyridinium chloride: 1,6-bis(2-pyridyl)hexane (1.62 g, 0.007 mol) was dissolved in 10 mL of DMF. The temperature was raised to 150 °C and dodecyl chloride (14.3 g, 0.07 mol) was slowly added to the solution. The reaction mixture was stirred for 18 h and the DMF was evaporated under reduced pressure. The residue was suspended in diethyl ether and crystallized two times from acetonitrile/toluene giving yellow crystals. The synthesis is reported in Scheme 1. Yield 2.64 g (58%); ^{1}H-NMR (400 MHz DMSO-d_6): δ 0.86 (t, J = 7.9 Hz, 6H); δ 1.23 br s, 40H); δ 1.72 (m, 4H); δ 1.84 (m, 4H); δ 3.09 (m, 4H); δ 4.56 (t, J = 8 Hz, 4H); δ 8.00 (m, 4H); δ 8.50 (t, J = 8 Hz, 2H); δ 9.03 (d, J = 8 Hz, 2H). ^{13}C NMR (100 MHz, DMSO-d_6): δ 158.2, 149.3, 136.7, 123.6, 121.3, 55.8, 44.2, 34.4, 32.7, 31.7, 29.5, 28.8, 26.7, 24.4, 22.6, 14.4, 13.8. FT-IR: ν = 3479, 3432, 2912, 2849, 1638, 1470, 1012, 933, 715 cm^{-1}. MS-ESI: m/z = 614 [M-Cl]. Anal. Calcd. For $C_{40}H_{70}Cl_2N_2$ (649.91): C 73.92, H 10.86, N 4.31. Found C 74.09, H 10.72, N 4.12.

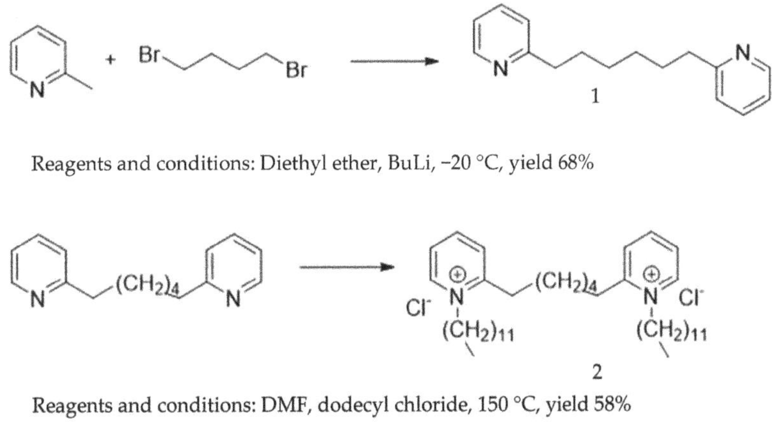

Reagents and conditions: Diethyl ether, BuLi, −20 °C, yield 68%

Reagents and conditions: DMF, dodecyl chloride, 150 °C, yield 58%

Scheme 1. Synthesis of 1,1′-bis-dodecyl-2,2′-hexane-1,6-diyl-bispyridinium chloride.

The NMR spectra were recorded using a Bruker 400 Avance (Billerica, MA, USA) spectrometer.
^{1}H NMR Spectra (400 MHz) chemical shifts (d scale) are reported in parts per million (ppm) and reported in order: multiplicity and number of protons; signals were characterized as s (singlet), d (doublet), t (triplet), m (multiplet), br s (broad signal).
^{13}C NMR (100 MHz); chemical shifts (d scale) are reported in parts per million (ppm).
IR spectra were recorded using an Agilent Technologies Cary 630 FTIR Spectrometer in the region 700–4000 without KBr support.
Mass spectra were recorded using an Applied Biosystem/MDS SCIEX API-150 EX instrument (Waltham, MA, USA).
The new compounds were analyzed on a ThermoQuest (Rodano, Italy) FlashEA 1112 Elemental Analyzer, for C, H, N. The percentages recorded were within 0.4% of the theoretical values.
The NMR, IR and mass spectra can be found in Supplementary Materials (Figures S1–S4).

3.3. ITC Measurements

ITC measurements of the dilution enthalpies of the surfactants under study were carried out on a MicroCal PEAQ-ITC (Malvern) at 25 °C. Surfactant solution was injected (first injection of 0.4 µL, followed by 18 injections of 2 µL each or 25 injections of 1.5 µL each) into a 200 µL reaction cell filled with doubly distilled and degassed water (as well as the reference cell) using a 40 µL automatized syringe with an interval of 150 or 120 s between two successive injections. A continuous stirring at 150 rpm was maintained throughout the experiments. The surfactant solutions used as titrants were prepared by weight using freshly boiled and degassed bi-distilled water at the concentrations of 10 and 20 mM. Data analysis was performed using MicroCal PEAQ-ITC Analysis Software (version 1.41, Malvern Panalytical, Malvern, UK).

3.4. Biological Assays

The biological assays of the new compound were carried out as described in ref. [18] to which we refer for experimental details. The human rhabdomyosarcoma cell line RD-4 (ATCC® CCL-136™) and the human pulmonary adenocarcinoma cell line A549 (ATCC® CCL-185™), cultured as described in ref. [18] were used in the experiments. Toxicity was evaluated with the MTT proliferation assay and the statistical differences among treatments were calculated using Student's test and multi-factorial ANOVA. Electrophoresis Mobility Shift Assay (EMSA) was used to evaluate the interaction between the cationic surfactant and the pEGFP-C1 plasmid, expressing the fluorescent green protein, prepared and stored as in ref. [18]. With a transient transfection assay, the gene delivery ability to the above cell lines of the compound under study was evaluated. Lipoplex formulations were prepared with the surfactant alone and by adding 1,2-dioleyl-sn-glycero-3-phosphoethanolamine (DOPE; SIGMA-Aldrich, St. Louis, CA, USA) to the plasmid–surfactant mixture at a surfactant:DOPE molar ratio of 1:2. Transfected cells were observed under a fluorescence microscope for EGFP expression.

3.5. Sample Preparation and AFM Imaging

A 20 µL droplet of a solution 0.1 nM of plasmid DNA in deposition buffer (4 mM Hepes, 10 mM NaCl, 2 mM $MgCl_2$, pH = 7.4), either in the presence or in the absence of the cationic lipid, was deposited onto freshly cleaved ruby mica (Ted Pella, Redding, CA, USA) after 5 min incubation at room temperature. The same N/P ratio, i.e., the ratio between the negative charges of the phosphate groups of DNA and the positive charges carried by the *gemini* pyridinium surfactant, as in transient transfection experiments, was used [18]. The mica disk was rinsed with Milli-Q water and dried with a weak nitrogen stream.

AFM imaging was carried out on the dried sample with a Park XE-100 microscope, operating in tapping mode, using commercial diving board silicon cantilevers (NSC-15 Micromash Corp., Sofia, Bulgaria). The software XEI (Park Systems, Suwon, Korea) was used for optimizing the images obtained by Park XE-100.

3.6. Circular Dichroism Spectroscopy

Circular dichroism (CD) experiments were performed using a Jasco 715 spectropolarimeter (JASCO International Co. Ltd., Tokyo, Japan), coupled with a Peltier PTC-348WI system for temperature control, set at 25 °C. CD spectra were the average of 4 scans recorded in the 220–320 nm range, using a 1 mm path length quartz cuvette, a bandwidth of 1 nm, data pitch of 0.5 nm, and a response time of 8 s.

Spectra were performed at a fixed pEGFP-C1 plasmid DNA concentration (1.36×10^{-3} M, with molarity expressed in terms of base pairs, bp) and in varying amounts of *gemini* surfactants to obtain the molar ratio 1:1, 1:2, 1:4 in 5 mM HEPES buffer, pH 7.0.

Following baseline correction, the measured ellipticity, θ (mdeg), was converted to the molar mean ellipticity [θ] (deg·cm^2·dmol^{-1}), using [θ] = θ/10 cl, where θ is ellipticity, c is the DNA molar concentration, and l is the optical path length in centimeters.

CD spectra deconvolution was performed using a gaussian model with Origin-Pro 2021.

3.7. Surface Tension Measurements

Surface tension measurements were performed using the Wilhelmy plate method, bringing a filter paper plate into contact with the surface of the liquid. The Wilhelmy plate was 0.75 square centimeters in area (0.5 × 1.5), attached to a balance with a thin metal wire. The force F on the plate due to wetting is measured with a NIMA PS4 tensiometer equipped with thermostatic support, to keep the temperature constant and avoid evaporation of water, and a trough in Teflon. The surface tension (γ) is calculated using the Wilhelmy equation:

$$\gamma = \frac{F}{l \, \cos(\theta)}$$

where l is the wetted perimeter of the Wilhelmy plate and θ is the contact angle between the liquid phase and the plate. In all the measurements reported here it was found $\theta = 0$, i.e., complete wetting conditions [38].

3.8. Ellipsometry

Ellipsometry measurements were performed using a multipurpose apparatus (Multiskop Optrel, Berlin, Germany) with a single wavelength of 632.8 nm. This technique was employed to evaluate the thickness of the surface layer, always in the Drude approximation [39]. The optical model adopted was that described in detail in ref. [40]. Film thickness was probed by the variation of the ellipsometric phase-angle $\delta\Delta$ which, in the Drude approximation, ref. [39], is linearly proportional to film thickness via a scaling factor depending on the incidence angle and refractive indexes. This specific apparatus provides an average thickness value calculated on an area of a few square millimetres.

4. Conclusions

Gene-based medicine is a clinical reality initially thought to fight against acquired or inherited genetic diseases and, after the pandemic spreading of the SARS-CoV-2 virus, well-known worldwide due to the use of vaccines based on viral DNA or mRNA. A great discussion is still open among scientists about the efficacy and the safety of this medical treatment, and research in this field is increasing more and more. Regardless of the kind of genetic material to be introduced inside the cells, it must be delivered by means of vectors, and the development of very efficient and safe non-viral vectors, avoiding immunogenic reaction and viral replication, is a landmark for in vivo gene-based medicine. We have devoted many years of our research to the synthesis, and chemico-physical and biological characterization, of *gemini* bis-pyridinium surfactants, both hydrogenated and partially fluorinated, and the results obtained confirm that pyridinium *gemini* surfactants, particularly those partially fluorinated, could be valuable tools for gene delivery purposes, but their performance highly depends on the spacer length and is strictly related to their structure in solution. To find out structure–activity relationships necessary to optimize their performance, we are collecting their chemico-physical properties in relation to their biological behaviour. The synthesis of the new hydrogenated compound having the spacer six methylene long and the hydrophobic tails of twelve carbon atoms each shows that the gene delivery ability of the hydrogenated compounds depends on the spacer length but barely on the hydrophobic tail length. Moreover, the physico-chemical data, particularly cmc and enthalpy of micellization, confirm that GP12_6 has a hydrophobicity very similar to FGP6, making their comparison correct. CD spectra have been shown to be useful tools to verify the formation of lipoplexes. In fact, under lipoplex formation, the CD spectra show a "tail" in the 288–320 nm region dependent on the presence of FGP6 and attributed to a chiroptical feature named ψ-phase. Such a phase is a consequence of the DNA structural transition from the B-form induced by the surfactant and reflects the supramolecular structure of the complex.

We have previously shown that the fluorinated compounds with spacer formed by 6 (FGP6) and 8 (FGP8) carbon atoms gives rise to a very interesting gene delivery activity, superior to that of the commercial reagent, when formulated with DOPE. Ellipsometric measurements suggest that FGP6 and FGP8 act in a very similar way, and dissimilar from FGP4, exactly as in the case of transfection, and confirm the hypothesis, suggested by thermodynamic data, about the requirement of a right length of the spacer for allowing the molecule to form a sort of molecular tong able to intercalate DNA. It remains to verify the effect of the hydrophobic chain length on fluorinated compounds, to optimize their efficiency.

Supplementary Materials: The following supporting information can be downloaded at: https://www.mdpi.com/article/10.3390/molecules28083585/s1. Figure S1: ^{13}C-NMR of 1,1′-bis-hexadecyl-2,2′-hexamethylenebispyridinium chloride (GP12_6).; Figure S2. ^{1}H-NMR of 1,1′-bis-dodecyl-2,2′-hexamethylenebispyridinium chloride (GP12_6). Figure S3. IR spectrum of 1,1′-bis-dodecyl-2,2′-hexamethylenebispyridinium chloride (GP12_6). Figure S4. Mass spectrum of 1,1′-bis-dodecyl-2,2′-hexamethylenebispyridinium chloride (GP12_6). Figure S5. Transfection of RD4 by GP12_6 as a function of concentration. On the left, phase contrast and, on the right, fluorescence microscope observation of the transfected cells (as shown by green cells expressing EGFP) are shown. The experiments were done with the surfactant alone and with surfactant:DOPE = 1:2. Cells are not transfected with DOPE alone. Figure S6. Transfection of A549 cells (right) by GP12_6 as a function of concentration. On the left, phase contrast and, on the right, fluorescence microscope observation of the transfected cells (as shown by green cells expressing EGFP) are shown. The experiments were done with the surfactant alone and with surfactant:DOPE = 1:2. Cells are not transfected with DOPE. Figure S7. Deconvolution of the spectra using Origin Pro, gaussian model. Figure S8. Kinetics of surface layer formation for FGP4 100 µM. It takes about 6 h to reach the equilibrium. Surface tension is lowered to 26 mN/m. Figure S9. Kinetics of surface layer formation for FGP4 50 µM. After 15 h, the equilibrium is not reached. Table S1. Deconvolution of the spectra. DNA: A-form yellow, B-form green. Table S2. Surface tension values of FGP4, FGP6 and FGP8 after 2 h. For each concentration, the mean and its relative positive and negative error are reported.

Author Contributions: Supervision, E.F.; Conceptualization, M.M. and E.F.; Methodology, M.R., C.C., T.A.P., D.O., V.F., G.D. and L.C.; Writing—original draft, M.M., E.F. and M.R.; Writing—review and editing, C.C., L.C., D.O., V.F., T.A.P. and G.D. All authors have read and agreed to the published version of the manuscript.

Funding: This research received no external funding.

Informed Consent Statement: Not applicable.

Data Availability Statement: All experimental data are reported in the paper and in Supplementary Materials.

Acknowledgments: The authors thank the Interdepartmental Centre of Measurements (CIM) of the University of Parma for allowing the use of the AFM facilities and the SCUSA Department for ITC instrumentation.

Conflicts of Interest: The authors declare no conflict of interest.

Sample Availability: Samples of the studied compounds are available from the authors.

References

1. Rosen, M.J. *Geminis*: A new generation of surfactants. *Chemtech* **1993**, *23*, 30–33.
2. Menger, F.M.; Keiper, J.S. *Gemini* surfactants. *Angew. Chem. Int. Ed.* **2000**, *39*, 1906–1920. [CrossRef]
3. Zana, R. *Novel Surfactants-Preparation, Applications, and Biodegradability*; Dimeric (*Gemini*) Surfactants in: Surfactant Science Series; Marcel Dekker, Inc.: New York, NY, USA, 1998; Volume 74, p. 241.
4. Zana, R.; Xia, J. *Gemini Surfactants, Interfacial and Solution Phase Behavior, and Applications*; Marcel Dekker, Inc.: New York, NY, USA, 2004.
5. Devínsky, F.; Pisárčik, M.; Lukáč, M. *Cationic Amphiphiles—Self-Assembling Systems for Biomedicine and Biopharmacy*; Nova Science Publishers, Inc.: New York, NY, USA, 2017.

6. Liu, L.; Wang, M.S.; Nair, J.; Yu, M.; Rapp, Q.; Wang, Y.; Luo, J.F.-W.; Chan, V.; Sahi, A.; Figueroa, X.V.; et al. Potent neutralizing antibodies against multiple epitopes on SARS-CoV-2 spike. *Nature* **2020**, *584*, 450–456. [CrossRef] [PubMed]
7. Okba, N.M.; Müller, M.A.; Li, W.; Wang, C.; Kessel, C.H.G.; Corman, V.M.; Lamers, M.M.; Sikkema, R.S.; Bruin, E.D.; Chandler, F.D.; et al. Severe acute respiratory syndrome coronavirus 2-specific antibody responses in coronavirus disease patients. *Emerg. Infect. Dis.* **2020**, *26*, 1478–1488. [CrossRef]
8. Sharma, V.; Ilies, M.A. Heterocyclic Cationic *Gemini* Surfactants: A Comparative Overview of Their Synthesis, Self-assembling, Physicochemical, and Biological Properties. *Med. Res. Rev.* **2014**, *34*, 1–44. [CrossRef]
9. Ahmed, T.; Kamel, A.O.; Wettig, S.D. Interactions between DNA and *Gemini* surfactant: Impact on gene therapy: Part I. *Nanomedicine* **2016**, *11*, 289–306. [CrossRef]
10. Fisicaro, E.; Compari, C.; Duce, E.; Donofrio, G.; Różycka-Roszak, B.; Wozniak, E. Biologically Active Bisquaternary Ammonium Chlorides: Physico-Chemical Properties of Long Chain Amphiphiles and Their Evaluation as Non-Viral Vectors for Gene Delivery. *Biochim. Biophys. Acta Gen. Subj.* **2005**, *1722*, 224–233. [CrossRef]
11. Viscardi, Q.G.; Barolo, C.; Barni, E.; Bellinvia, S.; Fisicaro, E.; Compari, C. *Gemini* Pyridinium Surfactants: Synthesis and Conductimetric Study of a Novel Class of Amphiphiles. *J. Org. Chem.* **2003**, *68*, 7651–7660. [CrossRef]
12. Fisicaro, E.; Compari, C.; Biemmi, M.; Duce, E.; Peroni, M.; Barbero, N.; Viscardi, G.; Quagliotto, P. The Unusual Behaviour of the Aqueous Solutions of *Gemini* Bispyridinium Surfactants: Apparent and Partial Molar Enthalpies of the Dimethanesulfonates. *J. Phys. Chem. B* **2008**, *112*, 12312–12317. [CrossRef]
13. Fisicaro, E.; Compari, C.; Bacciottini, F.; Barbero, N.; Viscardi, G.; Quagliotto, P. Is the Counterion Responsible for the Unusual Thermodynamic Behavior of the Aqueous Solutions of *Gemini* Bispyridinium Surfactants? *Colloids Surf. A Physicochem. Eng. Asp.* **2014**, *443*, 249–254. [CrossRef]
14. Fisicaro, E.; Compari, C.; Bacciottini, F.; Contardi, L.; Barbero, N.; Viscardi, G.; Quagliotto, P.; Donofrio, G.; Rozycka-Roszak, B.; Misiak, P.; et al. Nonviral Gene-Delivery: *Gemini* Bispyridinium Surfactant-Based DNA Nanoparticles. *J. Phys. Chem. B* **2014**, *118*, 13183–13191. [CrossRef]
15. Barbero, N.; Magistris, C.; Quagliotto, P.; Bonandini, L.; Barolo, C.; Buscaino, R.; Compari, C.; Contardi, L.; Fisicaro, E.; Viscardi, G. Synthesis and Physico-Chemical Characterization of Long Alkyl Chain *Gemini* Pyridinium Surfactants for Gene Delivery. *Chem. Plus Chem.* **2015**, *80*, 952–962. [CrossRef]
16. Quagliotto, P.; Barolo, C.; Barbero, N.; Barni, E.; Compari, C.; Fisicaro, E.; Viscardi, G. Synthesis and Characterization of Highly Fluorinated *Gemini* Pyridinium Surfactants. *Eur. J. Org. Chem.* **2009**, *19*, 3167–3177. [CrossRef]
17. Fisicaro, E.; Compari, C.; Bacciottini, F.; Contardi, L.; Pongiluppi, E.B.N.; Viscardi, G.; Quagliotto, P.; Donofrio, G.; Krafft, M. Nonviral Gene-Delivery by Highly Fluorinated *Gemini* Bispyridinium Surfactants-Based DNA Nanoparticles. *J. Colloid Interface Sci.* **2017**, *487*, 182–191. [CrossRef] [PubMed]
18. Massa, M.; Rivara, M.; Donofrio, G.; Cristofolini, L.; Peracchia, E.; Compari, C.; Bacciottini, F.; Orsi, D.; Franceschi, V.; Fisicaro, E. Gene-Delivery Ability of New Hydrogenated and Partially Fluorinated *Gemini* bispyridinium Surfactants with Six Methylene Spacers. *Int. J. Mol. Sci.* **2022**, *23*, 3062. [CrossRef]
19. Boussif, O.; Gaucheron, J.; Boulanger, C.; Santaella, C.; Kolbe, H.V.; Vierling, P. Enhanced in Vitro and in Vivo Cationic Lipid-Mediated Gene Delivery with a Fluorinated Glycerophosphoethanolamine Helper Lipid. *J. Gene Med.* **2001**, *3*, 109–114. [CrossRef] [PubMed]
20. Boulanger, C.; Di Giorgio, C.; Gaucheron, J.; Santaella, C. Transfection with Fluorinated Lipoplexes Based on New Fluorinated Cationic Lipids and in the Presence of a Bile Salt Surfactant. *Bioconjugate Chem.* **2004**, *15*, 901–908. [CrossRef] [PubMed]
21. Gaucheron, J.; Santaella, C.; Vierling, P. Improved in Vitro Gene Transfer Mediated by Fluorinated Lipoplexes in the Presence of a Bile Salt Surfactants. *J. Gen. Med.* **2001**, *3*, 338–344. [CrossRef]
22. Kissa, E. *Fluorinated Surfactants*; Marcel Dekker, Inc.: New York, NY, USA, 1994.
23. Riess, J.G. Highly Fluorinated Amphiphilic Molecules and Self-Assemblies with Biomedical Potential. *Curr. Opin. Colloid Interface Sci.* **2009**, *14*, 294–304. [CrossRef]
24. Krafft, M.; Riess, J. Selected physicochemical aspects of poly- and perfluoroalkylated substances relevant to performance. *Environ. Sustain. Part One Chemosphere* **2015**, *129*, 4–19. [PubMed]
25. Olofsson, G.; Loh, W. On the use of titration calorimetry to study the association of surfactants in aqueous solutions. *J. Braz. Chem. Soc.* **2009**, *20*, 577–593. [CrossRef]
26. Fisicaro, E.; Contardi, L.; Compari, C.; Bacciottini, F.; Pongiluppi, E.; Viscardi, G.; Barbero, N.; Quagliotto, P.; Różycka-Roszak, B. Solution Thermodynamics of Highly Fluorinated *Gemini* Bispyridinium Surfactants for Biomedical Applications. *Colloids Surf. A Physicochem. Eng. Asp.* **2016**, *507*, 236–242. [CrossRef]
27. McGhee, B.; Mingins, J.; Pethica, B.A. Thermodynamics of Micellization of Ionized Surfactants: The Equivalence of Enthalpies of Micellization from Calorimetry and the Variation of Critical Micelle Points with Temperature as Exemplified for Aqueous Solutions of an Aliphatic Cationic Surfactant. *Langmuir* **2021**, *37*, 8569–8576. [CrossRef]
28. Fisicaro, E.; Compari, C.; Duce, E.; Biemmi, M.; Peroni, M.; Braibanti, A. Thermodynamics of micelle formation in water, hydrophobic processes and surfactant self-assemblies. *Phys. Chem. Chem. Phys.* **2008**, *10*, 3903–3914. [CrossRef]
29. Shinoda, K.; Hato, M.; Hayashi, T. Physicochemical properties of aqueous solutions of fluorinated surfactants. *J. Phys. Chem.* **1972**, *76*, 909–914. [CrossRef]
30. Balgavý, P.; Devínsky, F. Cut-Off Effects in Biological Activities of Surfactants. *Adv. Colloid Interface Sci.* **1996**, *66*, 23–63. [CrossRef]

31. Jennings, K.H.; Marshall, I.C.; Wilkinson, M.J.; Kremer, A.; Kirby, A.J.; Camilleri, P. Aggregation Properties of a Novel Class of Cationic Gemini Surfactants Correlate with Their Efficiency as Gene Transfection Agent. *Langmuir* **2002**, *18*, 2426–2429. [CrossRef]
32. Neidle, S. *Nucleic Acid Structure and Recognition*; Oxford University Press: New York, NY, USA, 2002; pp. 89–138.
33. Grueso, E.; Cerrillos, C.; Hidalgo, J.; Lopez-Cornejo, P. Compaction and Decompaction of DNA Induced by the Cationic Surfactant CTAB. *Langmuir* **2012**, *28*, 10968–10979. [CrossRef]
34. Keller, D.; Bustamante, C. Theory of the interaction of light with large inhomogeneous molecular aggregates. II. Psi-type circular dichroism. *J. Chem. Phys.* **1986**, *84*, 2972–2980. [CrossRef]
35. Zhao, X.; Shang, Y.; Hu, J.; Liu, H.; Hu, Y. Biophysical characterization of complexation of DNA with oppositely charged *Gemini* surfactant 12-3-12. *Biophys Chem.* **2008**, *138*, 144–149. [CrossRef]
36. Bombelli, C.; Faggioli, F.; Luciani, P.; Mancini, G.; Sacco, M. Efficient transfection of DNA by liposomes formulated with cationic *gemini* amphiphiles. *J. Med. Chem.* **2005**, *48*, 5378–5382. [CrossRef] [PubMed]
37. Cristofolini, L. Synchrotron X-ray techniques for the investigation of structures and dynamics in interfacial systems. *Curr. Opin. Colloid Interface Sci.* **2014**, *19*, 228–241. [CrossRef]
38. Butt, H.J.; Graf, K.; Kappl, M. *Physics and Chemistry of Interfaces*; WILEY-VCH Verlag GmbH & Co. KGaA: Weinheim, Germany, 2003.
39. Tompkins, H.G.A. *User's Guide to Ellipsometry*; Academic Press: Boston, MA, USA, 1993.
40. Cristofolini, L.; Berzina, T.; Erokhina, S.; Konovalov, O.; Erokhin, V. Structural Study of the DNA Dipalmitoylphosphatidylcholine Complex at the Air-Water Interface. *Biomacromolecules* **2007**, *8*, 2270–2275. [CrossRef] [PubMed]

Disclaimer/Publisher's Note: The statements, opinions and data contained in all publications are solely those of the individual author(s) and contributor(s) and not of MDPI and/or the editor(s). MDPI and/or the editor(s) disclaim responsibility for any injury to people or property resulting from any ideas, methods, instructions or products referred to in the content.

Article

Interaction of Graphene Oxide Nanoparticles with Human Mesenchymal Stem Cells Visualized in the Cell-IQ System

Sergey Lazarev [1,2,*], Sofya Uzhviyuk [1], Mikhail Rayev [1,2], Valeria Timganova [1], Maria Bochkova [1,2], Olga Khaziakhmatova [3], Vladimir Malashchenko [3], Larisa Litvinova [3] and Svetlana Zamorina [1,2]

[1] Institute of Ecology and Genetics of Microorganisms, Ural Branch of the Russian Academy of Sciences-Branch of Perm Federal Research Center, 614081 Perm, Russia; kochurova.sofja@yandex.ru (S.U.); mbr_59@mail.ru (M.R.); timganovavp@gmail.com (V.T.); krasnykh-m@mail.ru (M.B.); mantissa7@mail.ru (S.Z.)

[2] Department of Microbiology and Immunology, Faculty of Biology, Perm State University, 614990 Perm, Russia

[3] Department of Microbiology and Immunology, Faculty of Biology, Immanuel Kant Baltic Federal University, 236041 Kaliningrad, Russia; hazik36@mail.ru (O.K.); vlmalashchenko@kantiana.ru (V.M.); larisalitvinova@yandex.ru (L.L.)

* Correspondence: lasest1999@gmail.com

Abstract: Graphene oxide is a promising nanomaterial with many potential applications. However, before it can be widely used in areas such as drug delivery and medical diagnostics, its influence on various cell populations in the human body must be studied to ensure its safety. We investigated the interaction of graphene oxide (GO) nanoparticles with human mesenchymal stem cells (hMSCs) in the Cell-IQ system, evaluating cell viability, mobility, and growth rate. GO nanoparticles of different sizes coated with linear or branched polyethylene glycol (P or bP, respectively) were used at concentrations of 5 and 25 µg/mL. Designations were the following: P-GOs (Ø 184 ± 73 nm), bP-GOs (Ø 287 ± 52 nm), P-GOb (Ø 569 ± 14 nm), and bP-GOb (Ø 1376 ± 48 nm). After incubating the cells with all types of nanoparticles for 24 h, the internalization of the nanoparticles by the cells was observed. We found that all GO nanoparticles used in this study exerted a cytotoxic effect on hMSCs when used at a high concentration (25 µg/mL), whereas at a low concentration (5 µg/mL) a cytotoxic effect was observed only for bP-GOb particles. We also found that P-GOs particles decreased cell mobility at a concentration of 25 µg/mL, whereas bP-GOb particles increased it. Larger particles (P-GOb and bP-GOb) increased the rate of movement of hMSCs regardless of concentration. There were no statistically significant differences in the growth rate of cells compared with the control group.

Keywords: graphene oxide nanoparticles; human mesenchymal stem cells; PEG; cell viability; cell growth; Cell-IQ; flow cytometry

Citation: Lazarev, S.; Uzhviyuk, S.; Rayev, M.; Timganova, V.; Bochkova, M.; Khaziakhmatova, O.; Malashchenko, V.; Litvinova, L.; Zamorina, S. Interaction of Graphene Oxide Nanoparticles with Human Mesenchymal Stem Cells Visualized in the Cell-IQ System. *Molecules* 2023, 28, 4148. https://doi.org/10.3390/molecules28104148

Academic Editors: Minas M. Stylianakis and Athanasios Skouras

Received: 25 April 2023
Revised: 11 May 2023
Accepted: 15 May 2023
Published: 17 May 2023

Copyright: © 2023 by the authors. Licensee MDPI, Basel, Switzerland. This article is an open access article distributed under the terms and conditions of the Creative Commons Attribution (CC BY) license (https://creativecommons.org/licenses/by/4.0/).

1. Introduction

Graphene is a gapless semiconductor that is currently being actively used in microelectronics and materials science [1–3] for various applications including the development of narrow-band electromagnetic absorbers [4] and metamaterials [5]. Due to the complexity of scalable production, its functional derivatives are more suitable for some applications. For example, graphene oxide (GO) offers a favorable alternative for use in biomedicine and optoelectronics due to its ease of fabrication, water solubility, and optical properties [6–9]. The graphene surface in GO is derivatized with epoxy, hydroxyl, and carboxyl groups, which enable it to form hydrogen bond-stabilized water suspensions [10–12]. In addition, GO has a large surface area available for functionalization and excellent mechanical properties [13,14], which make it attractive overall for electronics (LEDs and solar cells), tissue engineering, and drug delivery [15–18]. GO can be used as a basis for nanoscale sensors that detect small molecules such as NO_2 [19], proteins [20], influenza virus strains [21], and DNA strains [22]. It can also be used in pH measurement based on fluorescence to detect the microenvironment of tumors [23]. GO is effectively internalized by cells and shows stable fluorescence emission within cells as well as low cytotoxicity at the concentrations used for visualization [24,25]. This makes GO a potential candidate for drug delivery and in vitro or ex vivo visualization that can be used to detect and treat cancer [26]. However, in order for such applications of GO to become widely adopted, a solid understanding of the distribution of GO nanoparticles in vivo is required as well as knowledge of how it might interact with cells of different tissues.

Human mesenchymal stem cells (hMSCs), also known as multipotent mesenchymal stromal cells [27], are undifferentiated cells with the ability to self-renew and differentiate into various mesenchymal tissues, mainly bone, cartilage, and adipose tissue. According to the International Society for Cell Therapy, cells that possess the following characteristics can be considered human multipotent mesenchymal stromal cells [27]:

- Adhere to plastic;
- Differentiate in vitro into osteoblasts, adipocytes, and chondroblasts;
- Express CD105, CD90, CD73, CD44, and HLA-DR;
- Lack CD45, CD34, CD14, CD11b, CD79, and CD19.

Mesenchymal stem cells were first isolated from bone marrow in 1968. Since then, they have been isolated from many other tissues, including adipose tissue [28], perivascular networks [29], dental pulp [30], muscle, dermis, and embryonic tissue [31]. MSCs derived from bone marrow and adipose tissue are widely used in regenerative medicine [32]. It has been reported that drug-loaded hMSCs can deliver therapeutic cytokines to sites of injury or inflammation, making them a potentially useful tool in regenerative medicine and anti-tumor therapy [33]. Thus, hMSCs are attractive candidates for use as carriers of therapeutic agents and bioactive materials to specific target sites.

Cell-IQ® (CM-Technologies, Tampere, Finland) is a fully automated system for continuous in vitro cell imaging [34]. It includes an inverted phase contrast microscope with a built-in video camera and a climate chamber that maintains a constant temperature (37 °C) and CO_2 concentration (5%). Inside the climate chamber is a movable stand for a cell culture plate. Every 30 min, the system takes photographs in predetermined areas of the plate. The integrated software allows tracking of individual cells, including changes in their morphology and movement. The aim of this study was to investigate the interaction between GO nanoparticles and hMSCs with the Cell-IQ® system.

2. Results

2.1. Determination of the Relative Numbers of Live and Dead Cells after 24-h Incubation with GO Nanoparticles

Data on the number of live, apoptotic, and dead cells after 24-h incubation with GO nanoparticles, as determined by flow cytometry analysis, are shown in Figure 1. Cultivation of human mesenchymal stem cells in the presence of graphene oxide nanoparticles of types P-GOs, bP-GOs, and P-GOb at a concentration of 5 µg/mL did not result in a statistically significant decrease in the number of viable cells compared with the control group. The number of viable cells in these groups, as in the control group, was approximately 90%, which corresponds to the viability of the cells before the start of the experiment. At the same time, incubation of cells with these types of nanoparticles at a higher concentration of 25 µg/mL significantly decreased the relative number of viable cells. A decrease in the number of living cells correlated with an increase in the number of both dead and apoptotic cells, with a greater increase in the number of dead cells in all cases. The ratio between the number of dead cells and the number of apoptotic cells (D/A) in these samples was 1.74 ± 0.17, whereas in the control group D/A = 1.1.

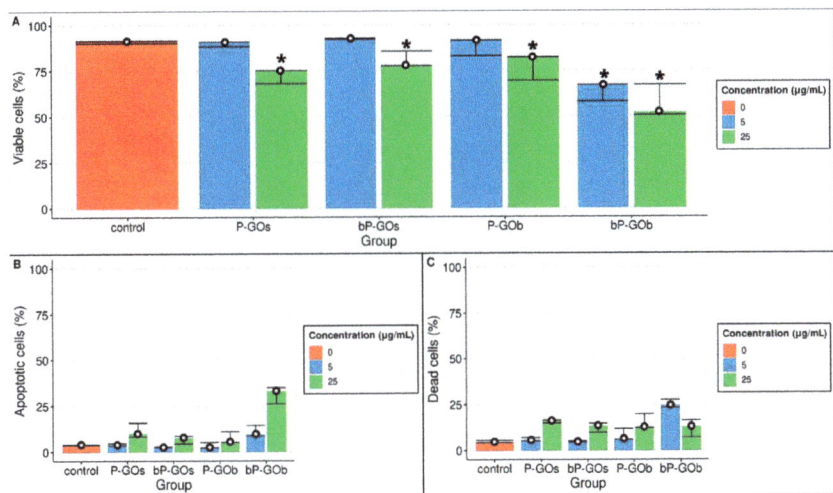

Figure 1. Comparison of the number of viable (**A**), apoptotic (**B**), and dead (**C**) cells after 24 h of cultivation in the presence of different types of GO nanoparticles (median, 1st and 3rd quartiles are shown). $n = 2$. Significant differences compared with control group (<0.05) are marked with *.

The greatest cytotoxicity among the samples studied was exerted by the bP-GOb nanoparticles. In this sample, cell viability decreased significantly after 24 h of cultivation, regardless of the concentration of nanoparticles used. Interestingly, the low concentration of bP-GOb resulted mainly in an increase in the number of dead cells (D/A = 2.54), whereas at a high concentration the loss of cell viability was mainly due to an increase in the number of apoptotic cells (D/A = 0.39).

Thus, it was shown that all particles studied had a cytotoxic effect on human mesenchymal stem cells at high concentrations. At low concentrations, the cytotoxic effect was observed only for bP-GOb particles. In all cases, there is a decrease in viability, mainly due to an increase in the number of dead cells. A decrease in viability due to the predominant increase in the number of apoptotic cells was observed only for the bP-GOb sample at high concentration.

2.2. Evaluation of Particle Internalization by Cells

After a 24-h incubation of cells with all samples of GO nanoparticles (except P-GOs at high concentration and bP-GOs at low concentration), a statistically significant increase in the number of highly granular cells was observed compared with the control group (Figure 2). This fact might indicate that either adhesion of the particles to the cell surface or their internalization occurs. We did not find any correlation between the concentration of particles and the percentage of high-granular cells, which may be due to several reasons. For example, particles that were weakly adsorbed on the cell surface might be lost during washing before flow cytometric analysis. In addition, we cannot exclude the influence of the physiological state of the cells (transition to apoptosis, death) on the granularity.

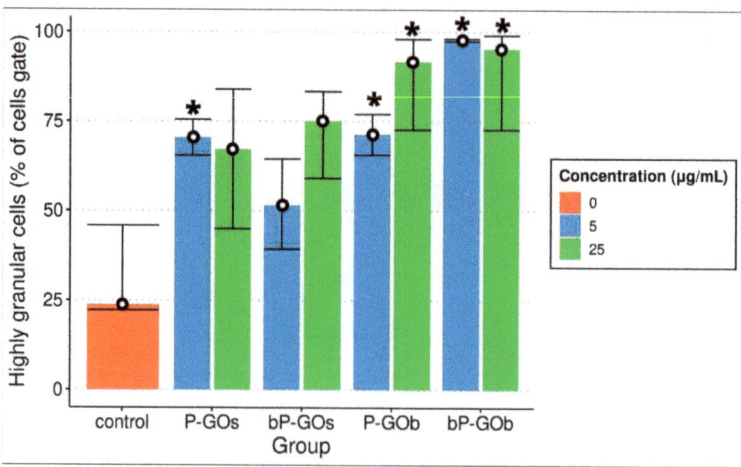

Figure 2. Comparison of the number of cells with high granularity after 24 h of co-cultivation with samples from GO (median is shown). $n = 2$. Significant differences compared with the control group ($p < 0.05$) are marked with *.

2.3. Evaluation of Cell Growth and Migration Activity in the Cell-IQ System

A total of 188 images were acquired for each well of the cell culture plate (47 images for each area of visualization) (Figure 3). Adhesion of particles on the cells is clearly visible in these images (Videos S1–S9). This is particularly noticeable in wells with high concentrations of samples. In addition, 24 h after the start of cultivation, areas without nanoparticles can be seen on the images. These correspond to the trajectories of hMSC movement as the cells have adhered/internalized most of the particles they encountered along the way.

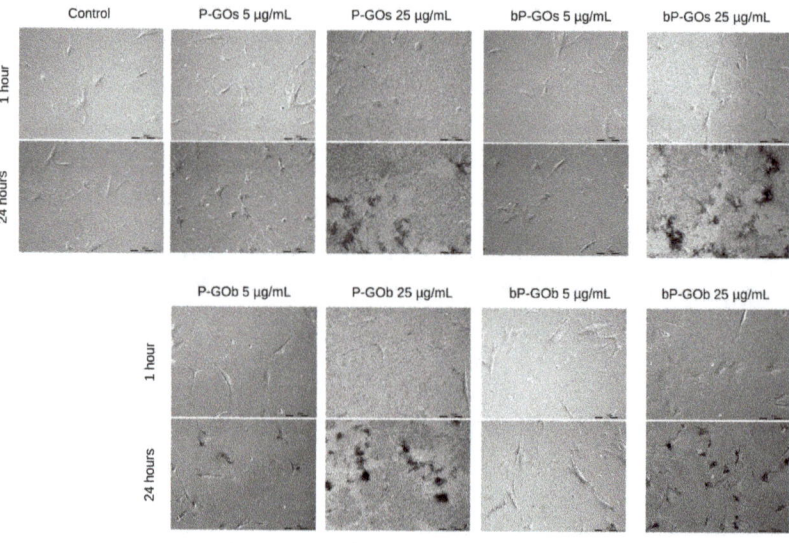

Figure 3. Photographs of cell culture plate wells taken in the Cell-IQ system. Each pair of images corresponds to the one area of visualization showing the changes during incubation of hMSCs with GO nanoparticles.

To assess the change in the number of cells in the visualization areas, the cells in every other image were counted. A pivot table was created based on the data obtained (Table S1). There were no statistically significant differences in either the increase in cell number after one day of cultivation or in the value of the K1 coefficient compared with the control group (Figure 4). In the evaluation of cell activity (K2), a statistically significant decrease in the activity of the cells cultured with the P-GOs sample at high concentration was observed compared to the control group. The opposite was true for the bP-GOb sample. At the minimum concentration, it apparently enhanced cell activity, resulting in a statistically significant increase in K2 levels compared to the control group. No statistically significant differences were found in the other samples compared to the control group. In general, it appears that low concentrations of GO nanoparticles contributed to an increase in cell activity, while high concentrations, on the contrary, had an inhibitory effect.

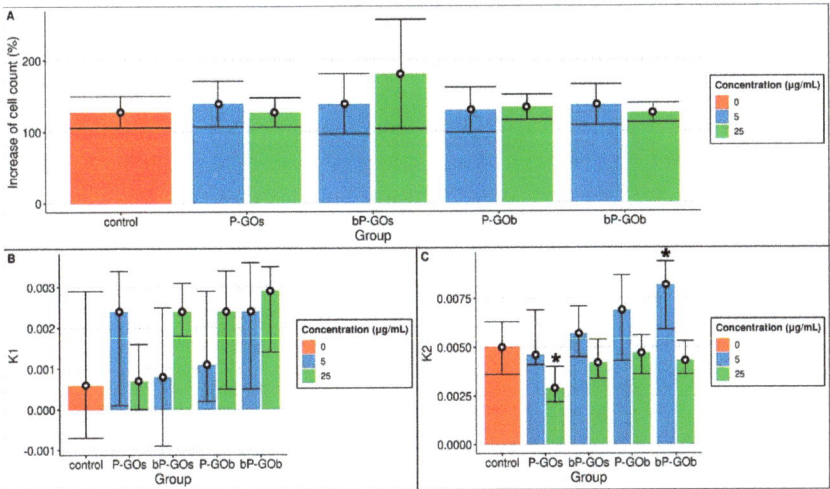

Figure 4. Comparison of (**A**) cell growth after 24 h of cultivation (mean and 95% confidence intervals shown); (**B**) change in cell number—K1 (median, 1st and 3rd quartiles shown). n = 12; (**C**) cell activity—K2 (median, 1st and 3rd quartiles shown). n = 12. Significant differences compared with control group ($p < 0.05$) are marked with *.

In addition, the velocity of cell movement was evaluated (Figure 5). An increase in the speed of cell movement was observed for all samples and concentrations studied compared to the control group, while this was statistically significant only for the P-GOb and bP-GOb samples.

To assess the effect of the samples on the functional properties of the hMSCs, the change in cell activity (K2) was evaluated. It can be seen that the concentration of the nanoparticles has a significant effect on the K2 value. The average K2 value for areas of visualization containing samples at a concentration of 25 µg/mL is lower than that of the control group, whereas it is higher for areas containing samples at a concentration of 5 µg/mL. Visually, it was observed that isolated cells moved farther and faster, while they slowed down when they were near other cells.

Overall, it was found that a high concentration of P-GOs particles statistically significantly reduced the value of the K2 coefficient, while high concentrations of bP-GOs particles increased K2. A statistically significant effect of other particle types on the K2 coefficient was not observed, but an upward trend for the K2 value can be seen for the samples in the following series: P-GOs → bP-GOs → P-GOb → bP-GOb (Figure 4). Moreover, a statistically significant increase in the speed of cell movement was observed for P-GOb and bP-GOb type particles at both concentrations, while no change in cell speed was registered for P-GOs and bP-GOs particles compared to the control.

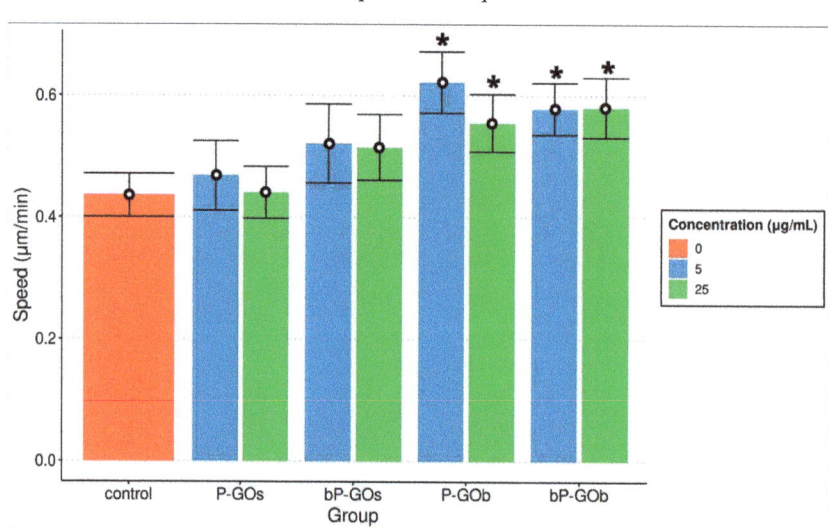

Figure 5. Comparison of the speed of cell movement during cultivation in the presence of different types of GO nanoparticles. Shown are the mean values and 95% confidence intervals of the mean values. n = 12. Significant differences compared with the control group ($p < 0.05$) are marked with *.

3. Discussion

For all samples studied, the adhesion/internalization of GO nanoparticles by the cells was visually observed in the images captured by the Cell-IQ system. This observation is confirmed by the number of highly granular cells determined by flow cytometry.

GO nanoparticles of the P-GOs, bP-GOs, and P-GOb types at a concentration of 5 µg/mL show a low level of cytotoxicity during short-term incubation with human mesenchymal cells, as there are no significant differences in the relative numbers of live and dead cells between the experimental and control groups. A high concentration of graphene oxide nanoparticles during short-term incubation results in a significant decrease in the number of live cells compared to the control group for all types of nanoparticles used in this study. Graphene oxide nanoparticles of the bP-GOb type have the most significant negative effect on the viability of human mesenchymal stem cells during short-term cultivation by significantly reducing the relative number of living cells, regardless of the concentration used.

Multiple mechanisms of cytotoxicity of GO nanoparticles have been described by other authors. The main mechanisms include interaction of cells with extremely sharp graphene edges [35], generation of reactive oxygen species (ROS) [36], and trapping of cells within aggregates of GO nanosheets [37]. In this study the mechanisms behind cytotoxicity of GO nanoparticles were not investigated and we possess no data regarding ROS generation or levels of oxidative stress the cells experienced. However, it is certain that the coating of GO nanoparticles by linear and especially branched PEG helped reduce the damage cells sustained from contact with sharp edges of graphene. Moreover, aggregation of nanoparticles on the cell surface was clearly seen by phase-contrast microscopy and may have contributed to cytotoxicity.

The nanoparticles used in this study had no significant effect on the growth of hMSCs, as reflected in both the K1 coefficient and the measurement of the final increase in cell number. Cell activity (K2) decreased significantly compared with the control group when cultivation was performed in the presence of a high-concentration P-GOs sample and increased when the concentration of the bP-GOs sample was low. Visually, a trend of increasing cell activity with increasing particle size can be observed.

In previous studies on the effects of other types of nanoparticles on MSCs, it has been repeatedly shown that these cells are able to internalize small particles (up to 200 nm) [38–40]. However, previous studies have predominantly used spherical polymer or metal particles. There are far fewer studies investigating the effect of graphene nanoparticles and their derivatives on MSCs. An important difference between graphene-based nanoparticles and those mentioned above is the two-dimensional structure of graphene-based materials, which provides a large surface area and the possibility of obtaining particles in different shapes.

The shape of the particles and surface modifications play an important role in achieving low cytotoxicity. For example, reduced GO nanoribbons were shown to cause more extensive DNA damage in hMSCs after a short incubation than nanosheets of similar size [41]. In another study using nanoparticles of PEG-coated reduced GO, incubation of cells for an extended period of time (up to 72 h) did not result in a significant decrease in viability [42]. In this case, particles with a diameter of 1 μm were used at concentrations of 5, 10, 50, and 100 μg/mL. The results obtained in our study are not consistent with these data. This could be due to several reasons, including the use of PEG with different lengths for nanoparticle coating. Syama et al. used PEG with a molecular weight of 1.9 kDa, whereas our particles were coated with PEG which had a molecular weight of 5 kDa.

The uptake of 2.7 μm-sized capsules by MSCs has previously been associated with a decrease in cell mobility [43]. The decrease in velocity was shown to be positively correlated with the concentration of capsules in the medium during incubation. At the same time, some studies have shown the ability of nanoparticles to accumulate in endosomes, leading to a change in cellular metabolism, stimulation of cells, and an increase in their mobility [44]. In this study, it was shown that incubation of cells with P-GOb and bP-GOb samples resulted in an increase in cell velocity, which may indicate that these particles have some effect on cellular metabolism, although the mechanism of this effect is not clear at present.

In general, there are good prospects for the use of graphene nanoparticles and their derivatives in diagnostics, therapy, and regenerative medicine. At the same time, the effect of these particles on human stem cells is still insufficiently studied. We see the need for further research in this area to gain a more comprehensive understanding of the parameters that play a key role in the cyto- and genotoxicity of graphene-based nanoparticles with respect to human stem cells. It is critical to know the consequences of short- or long-term exposure of human tissues to these materials.

4. Materials and Methods

Obtaining a cell culture. To study the effect of GO nanoparticles, a culture of human multipotent mesenchymal stem cells was obtained from a lipoaspirate at the Center for Immunology and Cellular Biotechnology at Immanuel Kant Baltic Federal University. The culture obtained met the minimum criteria set by the International Society for Cell Therapy [22]. The cells were stored in liquid nitrogen. After removal from the cryobank, the tube containing the frozen cell suspension was thawed in a water bath at 37 °C for 5 min. After thawing, the cells were washed with DMEM/F-12 containing 15 mM Hepes (Sigma-Aldrich, Saint Louis, MO, USA) by centrifugation at 1500 rpm for 5 min. Then, cells were transferred to T75 culture flasks (Eppendorf, Hamburg, Germany) and grown to 80% in complete culture medium (CCM) based on αMEM base medium (Sigma-Aldrich, Saint Louis, MO, USA) supplemented with 10% FBS (Sigma-Aldrich, Saint Louis, MO, USA), 100 U/mL penicillin, 100 μg/mL streptomycin (Thermo Fisher Scientific, Waltham, MA, USA), and 2 mM L-glutamine (Sigma-Aldrich, Saint Louis, MO, USA).

Properties of graphene oxide nanoparticles. Graphene oxide nanoparticles with sizes of 100–200 nm (GOs) and 1–5 μm (GOb) ("Ossila Ltd.", Sheffield, UK) coated with linear (P) and branched (bP) polyethylene glycol (PEG) were used. The procedures for modification and characterization of nanoparticles have been previously described by the authors [45]. The following particles were used in the study: P-GOs (Ø 184 ± 73 nm), bP-GOs (Ø 287 ± 52 nm), P-GOb (Ø 569 ± 14 nm), and bP-GOb (Ø 1376 ± 48 nm). The information about the types of nanoparticles used in this study is summarized in Table 1. Further information is available in [45] and Appendix B.

Table 1. Characteristics of GO nanoparticles.

Notation	Diameter (nm)	Coating
P-GOs	184 ± 73	Linear PEG
bP-GOs	287 ± 52	Branched PEG
P-GOb	569 ± 14	Linear PEG
bP-GOb	1376 ± 48	Branched PEG

Cultivation of cells for viability assessment. Cultivation was performed in CCM at 37 °C in a humidified atmosphere with 5% CO_2. Cells were cultured in a 6-well plate with 5 mL CCM per well. Cell suspension was added to the wells to a final concentration of $1 * 10^5$ cells per well. In addition, a suspension of GO nanoparticles of each type (P-GOs, bP-GOs, P-GOb, bP-GOb) was added to the wells at final concentrations of 5 and 25 µg/mL. Each group was present in duplicate. Cells without the addition of GO nanoparticles were used as negative controls. In addition, the cells with the addition of GO nanoparticles of each type at the maximum concentration (1 well for each sample type) were used as a null control.

Evaluation of cell culture viability. Analysis of cell viability before and after cultivation was performed by flow cytometry using MACS Quant FL7 (Miltenyi Biotec, Bergisch Gladbach, Germany). Guava ViaCount dye (Millipore, Burlington, MA, USA) was used for staining. This is a commercial reagent that combines an intercalating dye that selectively penetrates the membrane of dead cells and a specific dye for surface receptors, allowing accurate determination of the relative numbers of live, dead, and apoptotic cells.

Adherent culture of human mesenchymal stem cells was removed from the culture flasks by enzymatic treatment. The flask was washed 3 times with Hanks solution (HBSS) without magnesium and calcium (Capricorn Scientific, Ebsdorfergrund, Germany). Enzymatic treatment was performed with 3 mL of trypsin-EDTA solution (Sigma-Aldriche, Saint Louis, MO, USA). The gating strategy for flow cytometry is shown in Figure 6. Before the experiment, the viability of the cell culture was 90%. To compare the effect of the samples on apoptosis and cell death, the ratio between the number of dead cells and the number of apoptotic cells (D/A) was calculated.

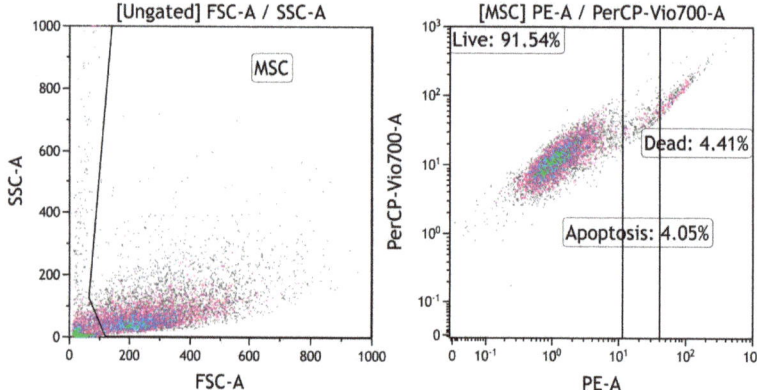

Figure 6. Gating strategy for human mesenchymal stem cells to assess the relative amounts of live, dead, and apoptotic cells.

Evaluation of nanoparticle adhesion/internalization by cells. In addition to viability, adhesion/internalization of particles by cells was also assessed. Cells that adhere/internalize GO nanoparticles became more granular, which increased the lateral scattering of light and could be detected by flow cytometry. To compare particle adhesion/internalization between samples, a subpopulation of high-granularity cells was isolated in the cell gate and the percentage of high-granularity cells was compared to that of the control group.

Cultivation of cells for assessment of their behavioral changes in the Cell-IQ system. Cultivation was performed in CCM at 37 °C in a humidified atmosphere with 5% CO_2 for 24 h. Cells were cultured in 24-well plates with 1 mL CCM per well. A suspension of cells was added to the wells of the cell culture plate at a concentration of $2*10^4$ cells per well. Then, a suspension of GO nanoparticles of each type was added to the wells to a final concentration of 5 and 25 µg/mL. Each group was present in triplicate. Cells cultured in CCM without addition of GO nanoparticles served as the control group.

For each well of the cell culture plate, four areas were selected for visualization (Figure 7). Images of these areas were taken every 30 min. The acquired images were analyzed to assess the activity of the cells. To obtain a better visual representation of the events taking place in each visualization area, the images for each of these areas were combined into a video file (one file per area, Videos S1–S9). As a result of image processing, the dynamics of cell growth and migration activity of human mesenchymal stem cells was evaluated.

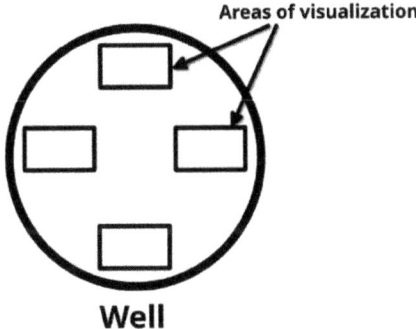

Figure 7. Distribution of visualization areas in a single well.

Evaluation of the dynamics of cell growth and migration activity of cells in the Cell-IQ system. Cells were counted in every second image acquired with the Cell-IQ system. Coefficients K1 and K2 were used to evaluate the dynamics of changes in cell number per area of visualization.

The K1 coefficient reflects changes in the number of cells in the observation area per unit time. It was calculated as the average value of the difference between the number of observed cells in each pair of analyzed images divided by the time interval between these images. In addition to K1, the final change in cell number in each visualization area was assessed. For this purpose, the ratio between the number of cells at the end of cultivation (after 24 h) and the number of cells at the beginning of cultivation was calculated for each area. This ratio reflects the total cell growth that occurred during cultivation.

$$K1 = \Sigma\left((k^i_{start} - k^i_{end})/t^i\right), \qquad (1)$$

k^i_{start}—number of cells observed in a single area of visualization at the beginning of a time period;

k^i_{end}—number of cells observed in a single area of visualization at the end of a time period;

t^i—time between observations.

The coefficient K2 reflects the activity of cell movement in the area of observation. It was calculated based on K1, but the absolute value of the difference between the number of cells observed in each pair of analyzed images was used. While K1 is influenced by cell growth, K2 is more appropriate to evaluate the migration of cells. If the average number of cells remains constant, their migration rate to/from the observation area can still be evaluated.

$$K2 = \Sigma \left(| k^i_{start} - k^i_{end} | / t^i \right), \qquad (2)$$

k^i_{start}—number of cells observed in a single area of visualization at the beginning of a time period;
k^i_{end}—number of cells observed in a single area of visualization at the end of a time period;
t^i—time between observations.

The speed of human mesenchymal stem cells was also evaluated. For this purpose, 3 cells were selected in each visualization area (if possible). Then, the central points of these cells in each image were determined. By tracking the changes in the coordinates of these points over time, the total distance travelled by each cell and its corresponding speed of movement could be calculated.

Statistical analysis. Descriptive statistics and hypothesis testing were performed using the standard STATISTICA package for Windows 10.0. Before data were analyzed, they were tested for normality using the Kolmogorov–Smirnov method. Median (Me), 25% (Q1), and 75% (Q3) quartiles were calculated for data that did not follow the normal distribution. Plots were generated with the R programming language [46] using the ggplot2 module [47].

For samples that did not follow the normal distribution, the significance of differences was assessed using the Wilcoxon T-test for pairs of dependent samples. Differences were considered significant when $p < 0.05$.

5. Conclusions

We studied the interaction of graphene oxide nanoparticles with hMSCs in the Cell-IQ in vivo monitoring system. We have shown that all the studied GO nanoparticles after 24 h of incubation have a cytotoxic effect on hMSCs at a high concentration (25 µg/mL), however, only bP-GOb particles were able to reduce cell viability at a low concentration (5 µg/mL).

We also showed that small particles (P-GOs) at a high concentration (25 µg/mL) reduced cell motility, while large-sized particles coated with branched PEG (bP-GOb) at a similar concentration increased this parameter. It was demonstrated that larger particles of P-GOb and bP-GOb (5 and 25 µg/mL) increased the speed of hMSCs movement.

However, the particles did not affect the changes in cell count during cultivation. In general, GO nanoparticles can reduce the viability of hMSCs, but increase the rate of cell movement. Thus, we have demonstrated for the first time the interaction of GO nanoparticles with hMSCs in the Cell-IQ in vivo monitoring system.

Supplementary Materials: The following supporting information can be downloaded at https://www.mdpi.com/article/10.3390/molecules28104148/s1: Video S1: Visualization of incubation of cells with P-GOs nanoparticles (5 µg/mL); Video S2: Visualization of incubation of cells with P-Gos nanoparticles (25 µg/mL); Video S3: Visualization of incubation of cells with bP-Gos nanoparticles (5 µg/mL); Video S4: Visualization of incubation of cells with bP-Gos nanoparticles (25 µg/mL); Video S5: Visualization of incubation of cells with P-Gob nanoparticles (5 µg/mL); Video S6: Visualization of incubation of cells with P-Gob nanoparticles (25 µg/mL); Video S7: Visualization of incubation of cells with bP-Gob nanoparticles (5 µg/mL); Video S8: Visualization of incubation of cells with bP-Gob nanoparticles (25 µg/mL); Video S9: Visualization of incubation of cells in the control group (without GO nanoparticles); Table S1: The pivot table of cell count in every area of visualization at different time periods.

Author Contributions: Conceptualization and funding acquisition, S.Z.; formal analysis, M.B.; investigation, S.U., O.K., M.B. and V.M.; project administration, M.R., L.L. and S.Z.; writing—original draft, S.L.; writing—review and editing, V.T. All authors have read and agreed to the published version of the manuscript.

Funding: This research was funded by the Russian Science Foundation, grant number 19-15-00244-П. The APC was funded by the same organization.

Institutional Review Board Statement: The study was conducted in accordance with the Declaration of Helsinki and approved by the Institutional Review Board (or Ethics Committee) of Immanuel Kant Baltic Federal University (Approval No. 1, 28 February 2019).

Informed Consent Statement: Informed consent was obtained from all subjects involved in the study.

Data Availability Statement: The datasets used and/or analyzed during the current study are available from the corresponding author on request.

Conflicts of Interest: The authors declare no conflict of interest. The funders had no role in the design of the study; in the collection, analyses, or interpretation of data; in the writing of the manuscript; or in the decision to publish the results.

Sample Availability: Samples of the GO nanoparticles used in this study are available from the authors.

Appendix A

Cultivation groups:

1. Control—cells in a culture medium without the addition of graphene oxide nanoparticles (3 iterations per plate, 6 iterations in total).
2. Sample 1 (5 µg/mL)—cells in a culture medium with the addition of P-GOs (Ø 184 ± 73 nm) graphene oxide nanoparticles at a concentration of 5 µg/mL (3 iterations).
3. Sample 1 (25 µg/mL)—cells in a culture medium with the addition of graphene oxide nanoparticles P-GOs (Ø 184 ± 73 nm) at a concentration of 25 µg/mL (3 iterations).
4. Sample 2 (5 µg/mL)—cells in a culture medium with the addition of graphene oxide nanoparticles bP-GOs (Ø 287 ± 52 nm) at a concentration of 5 µg/mL (3 iterations)
5. Sample 2 (25 µg/mL)—cells in a culture medium with the addition of graphene oxide nanoparticles bP-GOs (Ø 287 ± 52 nm) at a concentration of 25 µg/mL (3 iterations)
6. Sample 3 (5 µg/mL)—cells in a culture medium with the addition of graphene oxide nanoparticles P-GOb (Ø 569 ± 14 nm) at a concentration of 5 µg/mL (3 iterations)
7. Sample 3 (25 µg/mL)—cells in a culture medium with the addition of graphene oxide nanoparticles P-GOb (Ø 569 ± 14 nm) at a concentration of 25 µg/mL (3 iterations)
8. Sample 4 (5 µg/mL)—cells in a culture medium with the addition of graphene oxide nanoparticles bP-GOb (Ø 1376 ± 48 nm) at a concentration of 5 µg/mL (3 iterations)
9. Sample 4 (25 µg/mL)—cells in a culture medium with the addition of graphene oxide nanoparticles bP-GOb (Ø 1376 ± 48 nm) at a concentration of 25 µg/mL (3 iterations)
10. Group 0 control—in the second plate, row D is filled for unstained control. Cells in a culture medium with the addition of graphene oxide nanoparticles of each sample at a maximum concentration of 25 µg/mL.

Table A1. Layout of wells in experimental plates for analysis in the Cell-IQ in vitro imaging system.

	\multicolumn{6}{c}{Plate No. 1}
	1	2	3	4	5	6
A	Control 20,000 cells	Control 20,000 cells	Control 20,000 cells			
B	Sample 1 20,000 cells 5 μg/mL	Sample 1 20,000 cells 5 μg/mL	Sample 1 20,000 cells 5 μg/mL	Sample 2 20,000 cells 5 μg/mL	Sample 2 20,000 cells 5 μg/mL	Sample 2 20,000 cells 5 μg/mL
C	Sample 1 20,000 cells 25 μg/mL	Sample 1 20,000 cells 25 μg/mL	Sample 1 20,000 cells 25 μg/mL	Sample 2 20,000 cells 25 μg/mL	Sample 2 20,000 cells 25 μg/mL	Sample 2 20,000 cells 25 μg/mL
D						

	Plate No. 2					
	1	2	3	4	5	6
A	Control 20,000 cells	Control 20,000 cells	Control 20,000 cells			
B	Sample 3 20,000 cells 5 μg/mL	Sample 3 20,000 cells 5 μg/mL	Sample 3 20,000 cells 5 μg/mL	Sample 4 20,000 cells 5 μg/mL	Sample 4 20,000 cells 5 μg/mL	Sample 4 20,000 cells 5 μg/mL
C	Sample 3 20,000 cells 25 μg/mL	Sample 3 20,000 cells 25 μg/mL	Sample 3 20,000 cells 25 μg/mL	Sample 4 20,000 cells 25 μg/mL	Sample 4 20,000 cells 25 μg/mL	Sample 4 20,000 cells 25 μg/mL
D	Sample 1 20,000 cells 25 μg/mL	Sample 2 20,000 cells 25 μg/mL	Sample 3 20,000 cells 25 μg/mL	Sample 4 20,000 cells 25 μg/mL		

Table A2. Layout of wells in experimental plates for viability analysis by flow cytometry.

	Plate No. 1				Plate No. 2		
	1	2	3		1	2	3
A	Control 100,000 cells	Sample 1 100,000 cells 25 μg/mL	Sample 3 100,000 cells 25 μg/mL	A	Sample 1 100,000 cells 5 μg/mL	Sample 1 100,000 cells 25 μg/mL	Sample 2 100,000 cells 5 μg/mL
B	Control 100,000 cells	Sample 2 100,000 cells 25 μg/mL	Sample 4 100,000 cells 25 μg/mL	B	Sample 1 100,000 cells 5 μg/mL	Sample 1 100,000 cells 25 μg/mL	Sample 2 100,000 cells 5 μg/mL

	Plate No. 3				Plate No. 4		
	1	2	3		1	2	3
A	Sample 2 100,000 cells 25 μg/mL	Sample 3 100,000 cells 5 μg/mL	Sample 3 100,000 cells 25 μg/mL	A	Sample 4 20,000 cells 5 μg/mL	Sample 4 100,000 cells 25 μg/mL	Control 100,000 cells
B	Sample 2 100,000 cells 25 μg/mL	Sample 3 100,000 cells 5 μg/mL	Sample 3 100,000 cells 25 μg/mL	B	Sample 4 20,000 cells 5 μg/mL	Sample 4 100,000 cells 25 μg/mL	Control 100,000 cells

Appendix B

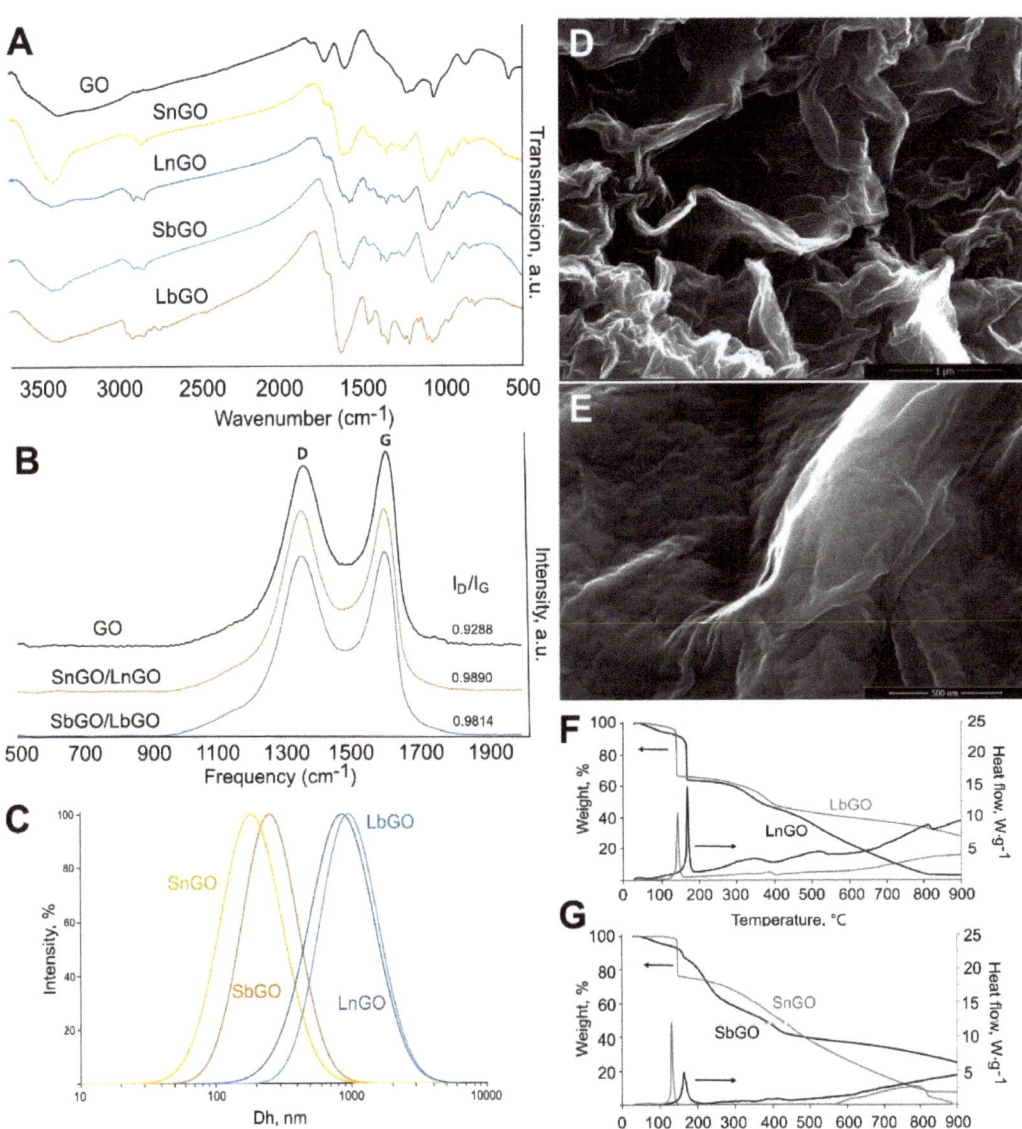

Figure A1. Characterization of P-GO nanoparticles (LnGO = P-GOb, LbGO = bP-GOb, SnGO = P-GOs, SbGO = bP-GOs). (**A**)—IR spectra; (**B**)—Raman spectra; (**C**)—intensity-weighted size distribution determined by DLS; (**D**,**E**)—SEM images of GO (**D**) and LbGO (**E**); (**F**,**G**)—TGA/DSC of P-GO. Scale bars are 1 μm (**D**) and 500 nm (**E**).

References

1. Schwierz, F. Graphene Transistors. *Nat. Nanotechnol.* **2010**, *5*, 487–496. [CrossRef] [PubMed]
2. Li, D.; Kaner, R.B. Graphene-Based Materials. *Science* **2008**, *320*, 1170–1171. [CrossRef] [PubMed]
3. Esteghamat, A.; Akhavan, O. Graphene as the Ultra-Transparent Conductive Layer in Developing the Nanotechnology-Based Flexible Smart Touchscreens. *Microelectron. Eng.* **2023**, *267–268*, 111899. [CrossRef]

4. Ye, Z.; Wu, P.; Wang, H.; Jiang, S.; Huang, M.; Lei, D.; Wu, F. Multimode Tunable Terahertz Absorber Based on a Quarter Graphene Disk Structure. *Results Phys.* **2023**, *48*, 106420. [CrossRef]
5. Lai, R.; Shi, P.; Yi, Z.; Li, H.; Yi, Y. Triple-Band Surface Plasmon Resonance Metamaterial Absorber Based on Open-Ended Prohibited Sign Type Monolayer Graphene. *Micromachines* **2023**, *14*, 953. [CrossRef]
6. Ishikawa, R.; Lugg, N.R.; Inoue, K.; Sawada, H.; Taniguchi, T.; Shibata, N.; Ikuhara, Y. Interfacial Atomic Structure of Twisted Few-Layer Graphene. *Sci. Rep.* **2016**, *6*, 21273. [CrossRef] [PubMed]
7. Luo, H.; Auchterlonie, G.; Zou, J. A Thermodynamic Structural Model of Graphene Oxide. *J. Appl. Phys.* **2017**, *122*, 145101. [CrossRef]
8. Chung, C.; Kim, Y.-K.; Shin, D.; Ryoo, S.-R.; Hong, B.H.; Min, D.-H. Biomedical Applications of Graphene and Graphene Oxide. *Acc. Chem. Res.* **2013**, *46*, 2211–2224. [CrossRef]
9. Ruan, Y.; Ding, L.; Duan, J.; Ebendorff-Heidepriem, H.; Monro, T.M. Integration of Conductive Reduced Graphene Oxide into Microstructured Optical Fibres for Optoelectronics Applications. *Sci. Rep.* **2016**, *6*, 21682. [CrossRef]
10. Mkhoyan, K.A.; Contryman, A.W.; Silcox, J.; Stewart, D.A.; Eda, G.; Mattevi, C.; Miller, S.; Chhowalla, M. Atomic and Electronic Structure of Graphene-Oxide. *Nano Lett.* **2009**, *9*, 1058–1063. [CrossRef]
11. Shin, D.S.; Kim, H.G.; Ahn, H.S.; Jeong, H.Y.; Kim, Y.-J.; Odkhuu, D.; Tsogbadrakh, N.; Lee, H.-B.-R.; Kim, B.H. Distribution of Oxygen Functional Groups of Graphene Oxide Obtained from Low-Temperature Atomic Layer Deposition of Titanium Oxide. *RSC Adv.* **2017**, *7*, 13979–13984. [CrossRef]
12. Matsumoto, Y.; Koinuma, M.; Taniguchi, T. Functional Group Engineering of Graphene Oxide. *Carbon* **2015**, *87*, 463. [CrossRef]
13. Montes-Navajas, P.; Asenjo, N.G.; Santamaría, R.; Menéndez, R.; Corma, A.; García, H. Surface Area Measurement of Graphene Oxide in Aqueous Solutions. *Langmuir* **2013**, *29*, 13443–13448. [CrossRef] [PubMed]
14. Papageorgiou, D.G.; Kinloch, I.A.; Young, R.J. Mechanical Properties of Graphene and Graphene-Based Nanocomposites. *Prog. Mater. Sci.* **2017**, *90*, 75–127. [CrossRef]
15. Wang, Y.; Li, Z.; Wang, J.; Li, J.; Lin, Y. Graphene and Graphene Oxide: Biofunctionalization and Applications in Biotechnology. *Trends Biotechnol.* **2011**, *29*, 205–212. [CrossRef]
16. Shadjou, N.; Hasanzadeh, M.; Khalilzadeh, B. Graphene Based Scaffolds on Bone Tissue Engineering. *Bioengineered* **2018**, *9*, 38–47. [CrossRef]
17. Mansilla Wettstein, C.; Bonafé, F.P.; Oviedo, M.B.; Sánchez, C.G. Optical Properties of Graphene Nanoflakes: Shape Matters. *J. Chem. Phys.* **2016**, *144*, 224305. [CrossRef]
18. Hu, W.; Li, Z.; Yang, J. Electronic and Optical Properties of Graphene and Graphitic ZnO Nanocomposite Structures. *J. Chem. Phys.* **2013**, *138*, 124706. [CrossRef]
19. Guo, L.; Hao, Y.-W.; Li, P.-L.; Song, J.-F.; Yang, R.-Z.; Fu, X.-Y.; Xie, S.-Y.; Zhao, J.; Zhang, Y.-L. Improved NO_2 Gas Sensing Properties of Graphene Oxide Reduced by Two-Beam-Laser Interference. *Sci. Rep.* **2018**, *8*, 4918. [CrossRef]
20. Huang, A.; Li, W.; Shi, S.; Yao, T. Quantitative Fluorescence Quenching on Antibody-Conjugated Graphene Oxide as a Platform for Protein Sensing. *Sci. Rep.* **2017**, *7*, 40772. [CrossRef]
21. Singh, R.; Hong, S.; Jang, J. Label-Free Detection of Influenza Viruses Using a Reduced Graphene Oxide-Based Electrochemical Immunosensor Integrated with a Microfluidic Platform. *Sci. Rep.* **2017**, *7*, 42771. [CrossRef] [PubMed]
22. Akhavan, O.; Ghaderi, E.; Rahighi, R. Toward Single-DNA Electrochemical Biosensing by Graphene Nanowalls. *ACS Nano* **2012**, *6*, 2904–2916. [CrossRef] [PubMed]
23. Campbell, E.; Hasan, M.T.; Pho, C.; Callaghan, K.; Akkaraju, G.R.; Naumov, A.V. Graphene Oxide as a Multifunctional Platform for Intracellular Delivery, Imaging, and Cancer Sensing. *Sci. Rep.* **2019**, *9*, 416. [CrossRef]
24. Ahmad, T.; Rhee, I.; Hong, S.; Chang, Y.; Lee, J. Ni-Fe_2O_4 Nanoparticles as Contrast Agents for Magnetic Resonance Imaging. *J. Nanosci. Nanotech.* **2011**, *11*, 5645–5650. [CrossRef] [PubMed]
25. Seabra, A.B.; Paula, A.J.; de Lima, R.; Alves, O.L.; Durán, N. Nanotoxicity of Graphene and Graphene Oxide. *Chem. Res. Toxicol.* **2014**, *27*, 159–168. [CrossRef] [PubMed]
26. Akhavan, O.; Ghaderi, E. Graphene Nanomesh Promises Extremely Efficient In Vivo Photothermal Therapy. *Small* **2013**, *9*, 3593–3601. [CrossRef]
27. Dominici, M.; Le Blanc, K.; Mueller, I.; Slaper-Cortenbach, I.; Marini, F.C.; Krause, D.S.; Deans, R.J.; Keating, A.; Prockop, D.J.; Horwitz, E.M. Minimal Criteria for Defining Multipotent Mesenchymal Stromal Cells. The International Society for Cellular Therapy Position Statement. *Cytotherapy* **2006**, *8*, 315–317. [CrossRef]
28. Zuk, P.A.; Zhu, M.; Ashjian, P.; De Ugarte, D.A.; Huang, J.I.; Mizuno, H.; Alfonso, Z.C.; Fraser, J.K.; Benhaim, P.; Hedrick, M.H. Human Adipose Tissue Is a Source of Multipotent Stem Cells. *MBoC* **2002**, *13*, 4279–4295. [CrossRef]
29. Crisan, M.; Yap, S.; Casteilla, L.; Chen, C.-W.; Corselli, M.; Park, T.S.; Andriolo, G.; Sun, B.; Zheng, B.; Zhang, L.; et al. A Perivascular Origin for Mesenchymal Stem Cells in Multiple Human Organs. *Cell Stem Cell* **2008**, *3*, 301–313. [CrossRef]
30. Gronthos, S.; Mankani, M.; Brahim, J.; Robey, P.G.; Shi, S. Postnatal Human Dental Pulp Stem Cells (DPSCs) in Vitro and in Vivo. *Proc. Natl. Acad. Sci. USA* **2000**, *97*, 13625–13630. [CrossRef]
31. Wang, H.; Hung, S.; Peng, S.; Huang, C.; Wei, H.; Guo, Y.; Fu, Y.; Lai, M.; Chen, C. Mesenchymal Stem Cells in the Wharton's Jelly of the Human Umbilical Cord. *Stem Cells* **2004**, *22*, 1330–1337. [CrossRef] [PubMed]
32. Bajek, A.; Gurtowska, N.; Olkowska, J.; Kazmierski, L.; Maj, M.; Drewa, T. Adipose-Derived Stem Cells as a Tool in Cell-Based Therapies. *Arch. Immunol. Ther. Exp.* **2016**, *64*, 443–454. [CrossRef] [PubMed]

33. Hu, Y.-L.; Fu, Y.-H.; Tabata, Y.; Gao, J.-Q. Mesenchymal Stem Cells: A Promising Targeted-Delivery Vehicle in Cancer Gene Therapy. *J. Control. Release* **2010**, *147*, 154–162. [CrossRef] [PubMed]
34. Narkilahti, S.; Rajala, K.; Pihlajamäki, H.; Suuronen, R.; Hovatta, O.; Skottman, H. Monitoring and Analysis of Dynamic Growth of Human Embryonic Stem Cells: Comparison of Automated Instrumentation and Conventional Culturing Methods. *BioMedical Eng. Online* **2007**, *6*, 11. [CrossRef]
35. Akhavan, O.; Ghaderi, E. Toxicity of Graphene and Graphene Oxide Nanowalls Against Bacteria. *ACS Nano* **2010**, *4*, 5731–5736. [CrossRef]
36. Dutta, T.; Sarkar, R.; Pakhira, B.; Ghosh, S.; Sarkar, R.; Barui, A.; Sarkar, S. ROS Generation by Reduced Graphene Oxide (RGO) Induced by Visible Light Showing Antibacterial Activity: Comparison with Graphene Oxide (GO). *RSC Adv.* **2015**, *5*, 80192–80195. [CrossRef]
37. Akhavan, O.; Ghaderi, E.; Esfandiar, A. Wrapping Bacteria by Graphene Nanosheets for Isolation from Environment, Reactivation by Sonication, and Inactivation by Near-Infrared Irradiation. *J. Phys. Chem. B* **2011**, *115*, 6279–6288. [CrossRef]
38. Ickrath, P.; Wagner, M.; Scherzad, A.; Gehrke, T.; Burghartz, M.; Hagen, R.; Radeloff, K.; Kleinsasser, N.; Hackenberg, S. Time-Dependent Toxic and Genotoxic Effects of Zinc Oxide Nanoparticles after Long-Term and Repetitive Exposure to Human Mesenchymal Stem Cells. *Int. J. Environ. Res. Public Health* **2017**, *14*, 1590. [CrossRef]
39. Brüstle, I.; Simmet, T.; Nienhaus, G.U.; Landfester, K.; Mailänder, V. Hematopoietic and Mesenchymal Stem Cells: Polymeric Nanoparticle Uptake and Lineage Differentiation. *Beilstein J. Nanotechnol.* **2015**, *6*, 383–395. [CrossRef]
40. Accomasso, L.; Rocchietti, E.C.; Raimondo, S.; Catalano, F.; Alberto, G.; Giannitti, A.; Minieri, V.; Turinetto, V.; Orlando, L.; Saviozzi, S.; et al. Fluorescent Silica Nanoparticles Improve Optical Imaging of Stem Cells Allowing Direct Discrimination between Live and Early-Stage Apoptotic Cells. *Small* **2012**, *8*, 3192–3200. [CrossRef]
41. Akhavan, O.; Ghaderi, E.; Emamy, H.; Akhavan, F. Genotoxicity of Graphene Nanoribbons in Human Mesenchymal Stem Cells. *Carbon* **2013**, *54*, 419–431. [CrossRef]
42. Syama, S.; Aby, C.P.; Maekawa, T.; Sakthikumar, D.; Mohanan, P.V. Nano-Bio Compatibility of PEGylated Reduced Graphene Oxide on Mesenchymal Stem Cells. *2D Mater.* **2017**, *4*, 025066. [CrossRef]
43. Litvinova, L.S.; Shupletsova, V.V.; Khaziakhmatova, O.G.; Daminova, A.G.; Kudryavtseva, V.L.; Yurova, K.A.; Malashchenko, V.V.; Todosenko, N.M.; Popova, V.; Litvinov, R.I.; et al. Human Mesenchymal Stem Cells as a Carrier for a Cell-Mediated Drug Delivery. *Front. Bioeng. Biotechnol.* **2022**, *10*, 796111. [CrossRef] [PubMed]
44. Vitale, E.; Rossin, D.; Perveen, S.; Miletto, I.; Lo Iacono, M.; Rastaldo, R.; Giachino, C. Silica Nanoparticle Internalization Improves Chemotactic Behaviour of Human Mesenchymal Stem Cells Acting on the SDF1α/CXCR4 Axis. *Biomedicines* **2022**, *10*, 336. [CrossRef] [PubMed]
45. Khramtsov, P.; Bochkova, M.; Timganova, V.; Nechaev, A.; Uzhviyuk, S.; Shardina, K.; Maslennikova, I.; Rayev, M.; Zamorina, S. Interaction of Graphene Oxide Modified with Linear and Branched PEG with Monocytes Isolated from Human Blood. *Nanomaterials* **2021**, *12*, 126. [CrossRef]
46. R Core Team. *R: A Language and Environment for Statistical Computing*; R Foundation for Statistical Computing: Vienna, Austria, 2021.
47. Wickham, H. *Ggplot2: Use R!* Springer International Publishing: Cham, Switzerland, 2016; ISBN 978-3-319-24275-0.

Disclaimer/Publisher's Note: The statements, opinions and data contained in all publications are solely those of the individual author(s) and contributor(s) and not of MDPI and/or the editor(s). MDPI and/or the editor(s) disclaim responsibility for any injury to people or property resulting from any ideas, methods, instructions or products referred to in the content.

Article

Efficient Preparation of Small-Sized Transition Metal Dichalcogenide Nanosheets by Polymer-Assisted Ball Milling

Qi Zhang, Fengjiao Xu, Pei Lu, Di Zhu, Lihui Yuwen * and Lianhui Wang *

State Key Laboratory of Organic Electronics and Information Displays & Jiangsu Key Laboratory for Biosensors, Institute of Advanced Materials (IAM), Nanjing University of Posts and Telecommunications, 9 Wenyuan Road, Nanjing 210023, China
* Correspondence: iamlhyuwen@njupt.edu.cn (L.Y.); iamlhwang@njupt.edu.cn (L.W.); Tel.: +86-25-85866333 (L.W.); Fax: +86-25-85866396 (L.W.)

Abstract: Two-dimensional (2D) transition metal dichalcogenide nanosheets (TMDC NSs) have attracted growing interest due to their unique structure and properties. Although various methods have been developed to prepare TMDC NSs, there is still a great need for a novel strategy combining simplicity, generality, and high efficiency. In this study, we developed a novel polymer-assisted ball milling method for the efficient preparation of TMDC NSs with small sizes. The use of polymers can enhance the interaction of milling balls and TMDC materials, facilitate the exfoliation process, and prevent the exfoliated nanosheets from aggregating. The WSe_2 NSs prepared by carboxymethyl cellulose sodium (CMC)-assisted ball milling have small lateral sizes (8~40 nm) with a high yield (~60%). The influence of the experimental conditions (polymer, milling time, and rotation speed) on the size and yield of the nanosheets was studied. Moreover, the present approach is also effective in producing other TMDC NSs, such as MoS_2, WS_2, and $MoSe_2$. This study demonstrates that polymer-assisted ball milling is a simple, general, and effective method for the preparation of small-sized TMDC NSs.

Keywords: high yield; polymer-assisted; ball milling; TMDC nanosheets

Citation: Zhang, Q.; Xu, F.; Lu, P.; Zhu, D.; Yuwen, L.; Wang, L. Efficient Preparation of Small-Sized Transition Metal Dichalcogenide Nanosheets by Polymer-Assisted Ball Milling. *Molecules* **2022**, *27*, 7810. https://doi.org/10.3390/molecules27227810

Academic Editors: Minas M. Stylianakis and Athanasios Skouras

Received: 1 November 2022
Accepted: 9 November 2022
Published: 12 November 2022

Publisher's Note: MDPI stays neutral with regard to jurisdictional claims in published maps and institutional affiliations.

Copyright: © 2022 by the authors. Licensee MDPI, Basel, Switzerland. This article is an open access article distributed under the terms and conditions of the Creative Commons Attribution (CC BY) license (https://creativecommons.org/licenses/by/4.0/).

1. Introduction

2D TMDCs have a unique layered structure in which the layers interact with each other by weak van der Waals force rather than by strong chemical bonding, which makes them different from other materials [1–4]. 2D TMDCs usually have the general formula of MX_2, in which transition metal (M) atoms and chalcogen (X) atoms form an X-M-X 'sandwich' structure, such as MoS_2, $MoSe_2$, WS_2, WSe_2, TiS_2, or $TiSe_2$ [1,5]. Since the interaction between TMDC layers is weak, few-layer or single-layer TMDC NSs can be obtained by physical or chemical exfoliation. These dimension changes in 2D TMDCs cause significant variations in their electronic structures and have a great impact on their physical and chemical properties [2,3]. For example, ultrathin MoS_2 NSs have high carrier mobility, enhanced photoluminescence, and high catalytic activity [6–9]. The size and aggregation of TMDC nanosheets also strongly influence the lubrication property in solid lubricants [10–12]. Moreover, when the lateral size of TMDC NSs is significantly reduced, typically at less than 20 nm, their properties are influenced by quantum confinement effects [13,14], which make small TMDC NSs promising nanomaterials for applications in optoelectronics, catalysis, energy storage, and biomedicine [3,5,15].

Over the last few decades, various methods have been developed to control the dimensions of TMDC NSs [2,4,16]. Bottom-up methods, such as chemical vapor deposition (CVD) and colloidal synthesis, can be used to synthesize TMDC NSs with various compositions and structures, for which high temperatures and rigorous reaction conditions are usually needed [17–19]. As well as bottom-up routes, top-down methods that involve exfoliating bulk TMDCs have also attracted great attention. Novoselov et al. developed a micromechanical exfoliation method for TMDC NSs by using Scotch tape [20]. Although high-quality TMDC NSs can be prepared by using this method, low yield and tedious work limit its use. Chemical exfoliation by alkaline metal intercalation is an efficient top-down approach to prepare ultrathin TMDC NSs [21–23]. However, the intercalation reactions usually need to be carried out in strict oxygen-free and water-free conditions, and the as-prepared TMDC NSs often have altered crystal structures with the formation of abundant defects. Ultrasonication-assisted liquid phase exfoliation (LPE) is another popular method that has been extensively explored for the preparation of TMDC NSs due to its simplicity and generality [24,25]. However, the efficiency of LPE is relatively low, especially for the preparation of TMDC NSs with small sizes [24–26]. Ball milling is another top-down method to produce TMDC NSs with the potential for large-scale production [27–30]. Compared with the normal force-dominated LPE method, ball milling utilizes both shear force and normal force provided by the milling balls to exfoliate and pulverize the layered TMDCs [31,32]. Similar to the LPE method, the low yield of nanosheets is also a common issue due to the limited milling ball–material contact interface and the reaggregation of nanosheets during ball milling. Although sonication-assisted exfoliation has been used together with ball milling, the improvement of the yield is still limited and the operation becomes complicated [28,33,34]. Therefore, a novel method for highly efficient preparation of small TMDC NSs with high yield and simplicity is still needed.

In this report, we develop a polymer-assisted ball milling strategy for the efficient preparation of TMDC NSs with small sizes (Scheme 1). By using carboxymethyl cellulose sodium (CMC) as a solid intermedium in the ball milling process, WSe_2 NSs with different average sizes from about 8 nm to 40 nm were successfully prepared with a total yield of over 60%. The influence of the polymer, rotation speed, and milling time on the size and yield of WSe_2 NSs was studied. This method provides polymer surface modification during preparation, which gives the as-prepared WSe_2 NSs good colloidal stability. Moreover, this method can also be used to prepare other TMDC NSs, such as MoS_2, $MoSe_2$, and WS_2. Our study demonstrates a novel method for the preparation of small-sized TMDC NSs with high yields by using polymer-assisted ball milling, which is efficient, simple, general, and scalable.

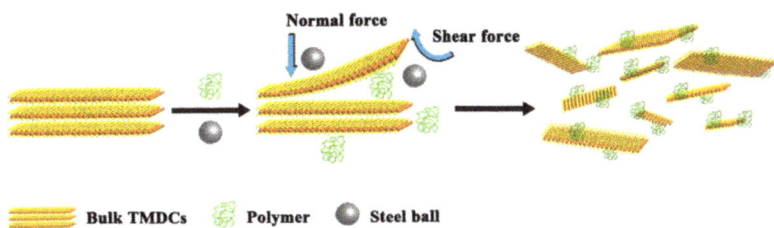

Scheme 1. Preparation of small TMDC NSs by polymer-assisted ball milling.

2. Results and Discussion

2.1. WSe₂ NSs Prepared by CMC-Assisted Ball Milling

As illustrated in Scheme 1, polymers are mixed with bulk TMDC materials in a steel jar to assist the dry ball milling process. The shear force provided by the steel balls leads to the exfoliation of the layered materials, while the normal force fragments the nanosheets. During ball milling, the polymers act as an intermedium to transfer the impact from steel balls to TMDC materials and enhance the interaction between steel balls and TMDC materials. Moreover, the polymer molecules can adsorb on the surface of TMDC NSs and reduce their reaggregation, which also improves the exfoliation efficiency. Polymer-assisted ball milling not only decreases the thickness and lateral size of layered TMDCs but also produces homogeneous TMDC–polymer composites. Without further ultrasonication treatment, the as-prepared TMDC NSs can easily disperse in water and form stable colloidal dispersions. TMDC NS aqueous dispersions were first centrifuged at low speed to remove the large aggregates, and the supernatant was further centrifuged at different speeds to obtain TMDC NSs with different sizes.

CMC was first explored to prepare WSe_2 NSs by ball milling. As a carboxymethyl functionalized cellulose, CMC is an abundant, cheap, and environmentally friendly material and has been extensively used in various industrial applications [35]. After ball milling, CMC-WSe_2 NSs composites can easily form aqueous dispersions due to the electrostatic repulsion force provided by CMC. After gradient centrifugation at different rotation speeds, a series of small WSe_2 NSs were obtained and named as WSe_2-Low (10,000 rpm), WSe_2-Medium (16,000 rpm), and WSe_2-High (21,000 rpm), respectively. TEM was used to investigate the morphology of WSe_2 NSs. Figure 1a–c and Figure S1 show that WSe_2 NSs have uniform sheet-like morphology with average sizes of 39.66 ± 13.63 nm (WSe_2-Low), 20.02 ± 6.28 nm (WSe_2-Medium), and 7.90 ± 3.11 nm (WSe_2-High), respectively. As shown by the HRTEM images in Figure 1d–f, WSe_2 NSs have a clear crystalline structure and the lattice spacing of 2.8 Å can be assigned to the (100) plane of 2H-WSe_2 [36]. The six-fold SAED patterns (Figure 1g–i) indicate that the WSe_2 NSs with different sizes have the same hexagonal symmetry structure, suggesting no significant change in the structure during the ball milling process.

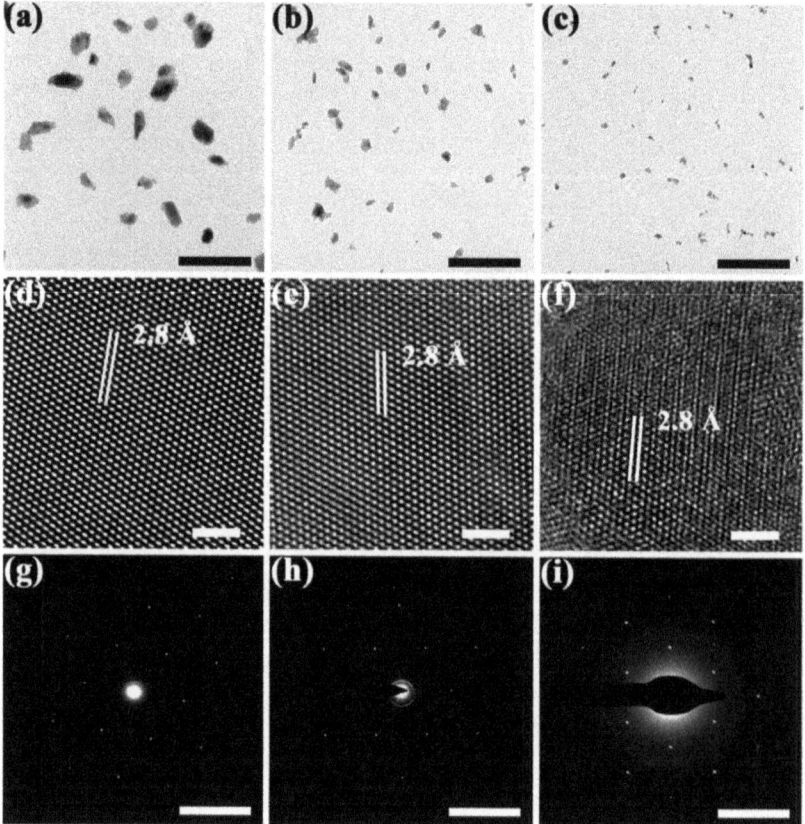

Figure 1. (**a**–**c**) TEM, (**d**–**f**) HRTEM, and (**g**–**i**) SAED images of WSe_2 NSs with different sizes prepared by CMC–assisted ball milling at different centrifugation speeds: (**a**,**d**,**g**) WSe_2–Low (10,000 rpm for 1 h), (**b**,**e**,**h**) WSe_2–Medium (16,000 rpm for 1.5 h), (**c**,**f**,**i**) WSe_2–High (21,000 rpm for 4 h). Scale bars: 200 nm for (**a**–**c**), 2 nm for (**d**–**f**), and 5 1/nm for (**g**–**i**).

Atomic force microscopy (AFM) was used to investigate the thickness of WSe$_2$ NSs. AFM images (Figures 2 and S2) show that the average thickness of WSe$_2$ NSs gradually decreases from 8–10 nm (WSe$_2$-Low) to 2–3 nm (WSe$_2$-High) according to the increase in centrifugation speed from 10,000 rpm to 21,000 rpm. Hence, small-sized WSe$_2$ NSs with different thicknesses can be obtained by varying the rotation speed during gradient centrifugation. Since CMC molecules can still absorb on the surface of WSe$_2$ NSs after purification, the apparent thickness of WSe$_2$ NSs detected by AFM may be a little higher than the exact true values.

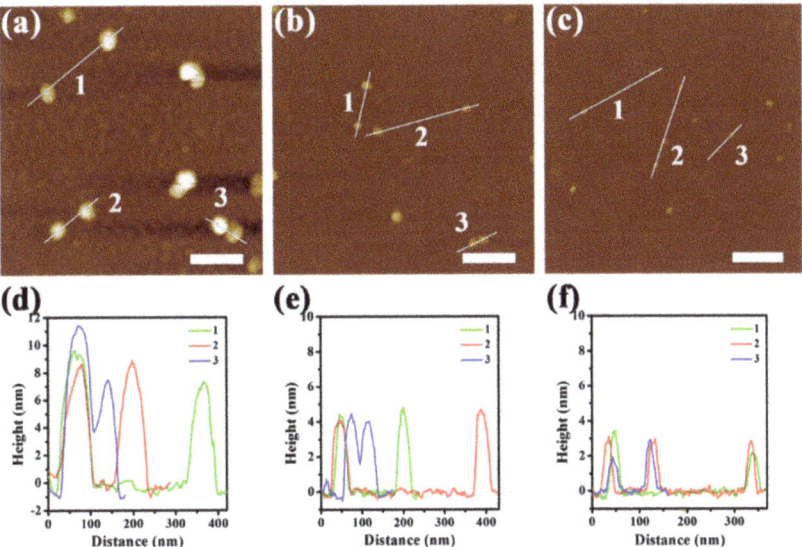

Figure 2. (**a**–**c**) AFM images and (**d**–**f**) corresponding height profiles of WSe$_2$ NSs with different sizes prepared by CMC–assisted ball milling after gradient centrifugation: (**a**,**d**) Wse$_2$–Low, (**b**,**e**) Wse$_2$–Medium, (**c**,**f**) Wse$_2$–High. Scale bar: 200 nm.

The structure, composition, and properties of the WSe$_2$ NSs were further studied. As shown by the XRD patterns in Figure 3a, diffraction peaks at 13.6°, 31.4°, 37.8°, and 47.4° belong to the (002), (100), (103), and (105) planes according to the standard diffraction data of WSe$_2$ (JCPDS, 38-1388). The broadening and weakening of these diffraction lines originate from the size and thickness reduction of WSe$_2$ NSs after ball milling [37,38]. XPS was used to investigate the chemical state of the as-prepared WSe$_2$ NSs. As shown by the core-level XPS spectra of W in Figure 3b, the doublet peaks near 33 eV and 35 eV belong to W^{4+} 4f$_{7/2}$ and W^{4+} 4f$_{5/2}$ of 2H-WSe$_2$, while the binding energy peak located at about 38 eV can be ascribed to W^{4+} 5p$_{3/2}$ [39]. The absence of doublet peaks for W^{6+} between 36 and 38 eV indicates that no WO$_x$ formed during ball milling [40]. As illustrated in Figure 3c, the doublet peaks near 55 eV and 56 eV belong to Se 3d$_{5/2}$ and Se 3d$_{3/2}$ of 2H-WSe$_2$ [41]. Similarly, WSe$_2$ NSs with other sizes (WSe$_2$-Low and WSe$_2$-High) have the same chemical states of W and Se as WSe$_2$ Medium (Supplementary Materials Figure S3), which implies that WSe$_2$ NSs were prepared by polymer-assisted ball milling mainly through physical exfoliation and fragmentation rather than the mechanochemical way. Raman spectra of WSe$_2$ NSs (Figure 3d) show that the characteristic peak located at about 250 cm^{-1} can be ascribed to the degenerate A$_{1g}$ (out-of-plane) and E$^1_{2g}$ (in-plane) vibrational modes, suggesting the few-layer structure of WSe$_2$ NSs [39,41]. UV-vis-NIR spectra of WSe$_2$ NSs (Figure 3e) indicate that the absorption peaks near 750 nm blueshift with the size and thickness reduction, similar to previous reports [42,43]. As the FT-IR spectra of WSe$_2$ NSs depict in Figure 3d, IR absorption bands at 3430 cm^{-1} and 1630 cm^{-1} can be assigned to the stretching vibration of the hydroxyl group (−OH) and asymmetric vibration of the carboxyl group (COO−) of CMC [44,45], respectively, which suggests the existence of CMC molecules on the surface of WSe$_2$ NSs. Due to the hydrophilicity and electrostatic repulsion of CMC, WSe$_2$ NSs are stable in phosphate buffer saline (PBS) and cell culture medium (DMEM) (Supplementary Materials Figure S4), and no obvious precipitation can be observed even after being stored for months in water, which suggests their good colloidal stability in biological environments.

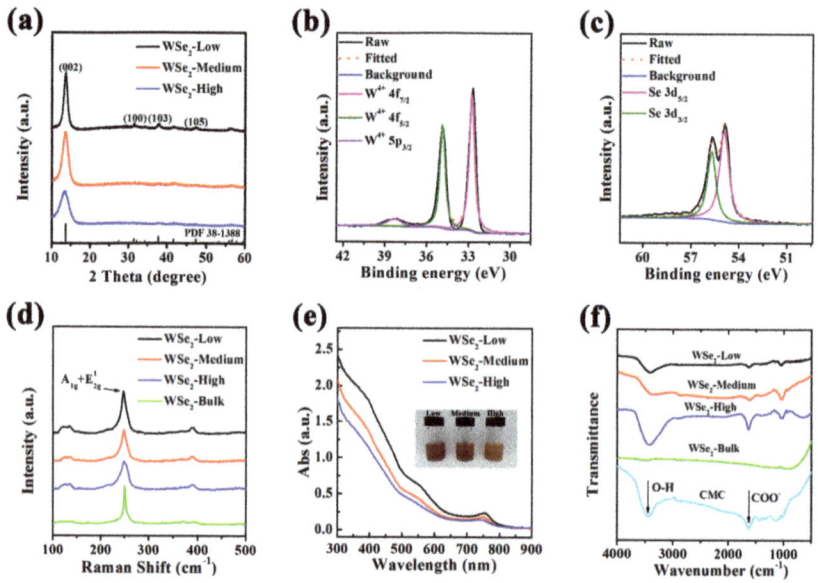

Figure 3. (a) XRD patterns of WSe$_2$ NSs with different sizes. XPS spectra of WSe$_2$ NSs (WSe$_2$–Low) for (b) W 4f and W 5p, and (c) Se 3d. (d) Raman spectra of bulk WSe$_2$ and WSe$_2$ NSs with different sizes. (e) UV–vis–NIR absorption spectra and photos of WSe$_2$ NSs aqueous dispersions (inset). (f) FT-IR spectra of WSe$_2$ NSs, bulk WSe$_2$, and CMC.

2.2. The Influence of Experimental Conditions on Polymer-Assisted Ball Milling

The influence of the experimental conditions of ball milling on the size and yield of WSe$_2$ NSs was studied. Due to the important role of the polymer during ball milling, different polymers were first studied. As shown in Supplementary Materials Figure S5, WSe$_2$ NSs prepared by using different polymers all have uniform morphology and good dispersity with an average size smaller than 100 nm. Table 1 indicates that the total yield of WSe$_2$ NSs by CMC-assisted ball milling is 62.68%, while the yields of WSe$_2$ NSs using F127, PVP, and PEG are 12.34%, 32.08%, and 11.89%, respectively. Moreover, the lateral size of WSe$_2$ NSs using CMC is much smaller than those using other polymers. The superiority of CMC for the ball milling process may originate from its unique structure. Previous studies have revealed that the hydroxyl and carboxyl groups of polymers, such as alginate, bovine serum albumin, and glycan, have a strong interaction with the surface of TMDC NSs, and the synergy of the repetitive units of the polymers also enhances this interaction [46–48]. Compared with F127, PVP, and PEG, CMC has much more hydroxyl and carboxyl groups, which may provide a stronger interaction between the surface of WSe$_2$ NSs and polymers and significantly improve the efficiency of ball milling. Milling time and rotation speed during ball milling also play important roles in the morphology and yield of the WSe$_2$ NSs. As shown in Table 1 and Supplementary Materials Figure S6, the lateral size of WSe$_2$ NSs generally reduces with increased milling time, while total yields of WSe$_2$ NSs are nearly the same (~60%), suggesting that the milling time mainly influences the size of WSe$_2$ NSs rather than the yield. As shown in Supplementary Materials Figure S7 and Table 1, the size of most WSe$_2$ NSs decreases from about 60 nm to 30 nm when the rotation speed of the ball mill increases from 400 rpm to 800 rpm. Meanwhile, the total yield of WSe$_2$ NSs increases from about 30% to 60% once the rotation speed has exceeded 650 rpm. Therefore, both the size and yield of WSe$_2$ NSs can be easily adjusted by varying the rotation speed and milling time during ball milling, suggesting the controllability of this method.

Table 1. Average lateral size and yield of WSe$_2$ NSs prepared by polymer-assisted ball milling under different experimental conditions.

Polymer	Speed (rpm)	Time (h)	Size (nm)/Yield (%)			Total Yield (%)
			WSe$_2$-Low	WSe$_2$-Medium	WSe$_2$-High	
CMC	650	12	39.66/26.32	20.02/26.18	7.90/10.18	62.68
F127	650	12	93.20/7.68	72.39/3.84	48.73/0.82	12.34
PVP	650	12	46.29/23.48	19.05/8.39	14.87/0.21	32.08
PEG	650	12	47.05/8.81	37.00/2.21	30.76/0.87	11.89
CMC	650	6	55.04/22.34	35.28/25.60	21.54/14.36	62.30
CMC	650	24	24.06/32.52	12.41/23.72	7.68/3.82	60.06
CMC	400	12	61.05/26.92	51.44/1.32	35.78/0.56	28.80
CMC	800	12	28.26/23.61	15.82/21.34	7.46/13.92	58.87

2.3. The Preparation of other TMDC NSs by Polymer-Assisted Ball Milling

The feasibility of polymer-assisted ball milling for the preparation of other TMDC NSs was investigated. As shown in Figure 4a–c, the MoS$_2$, MoSe$_2$, and WS$_2$ NSs prepared by CMC-assisted ball milling have uniform morphology and good dispersity. These TMDC NSs obtained by medium centrifugation speed (16,000 rpm) have lateral sizes similar to WSe$_2$-Medium (~20 nm) (Figure S8, Supplementary Materials), suggesting the general applicability of this method. HRTEM images (Figure 4d–f) show that MoS$_2$, MoSe$_2$, and WS$_2$ NSs all have a clear crystalline structure with identical lattice spacings to previous reports [36,46]. The SAED patterns (Figure 4g–i) further indicate the hexagonal symmetry structure of these TMDC NSs, which is characteristic of the crystal structure of the 2H phase [24,36]. As shown in Figure 5 and Supplementary Materials Figure S9, the average thickness of these TMDC NSs is in the range of 4–6 nm, suggesting they have a similar few-layer structure to WSe$_2$ NSs.

To further confirm the successful preparation of the MoS$_2$, MoSe$_2$, and WS$_2$ NSs, structure, composition, and property characterizations were performed. As illustrated in Figure 6a, diffraction peaks at 14.3° (MoS$_2$), 13.7° (MoSe$_2$), and 14.3° (WS$_2$) belong to the (002) plane, respectively, similar to the standard diffraction data. The weakening of (100), (103), and (105) lines may originate from the size and thickness reduction of these TMDC NSs [37]. The doublet peaks in the XPS spectrum of Mo (Figure 6b) near 229 eV and 232 eV correspond to Mo^{4+} 3d$_{5/2}$ and Mo^{4+} 3d$_{3/2}$ of 2H-MoS$_2$, respectively; and the weak peak near 227 eV corresponds to S 2s [34,49]. The absence of the peak near 237 eV for Mo^{6+} 3d suggests no oxidation of MoS$_2$ [40]. The doublet peaks near 162 eV and 163 eV in Figure 6c can be ascribed to the S 2p$_{3/2}$ and S 2p$_{1/2}$ of 2H-MoS$_2$, respectively. In addition, the high-resolution XPS spectra (Supplementary Materials Figure S10) further confirm the successful preparation of 2H phase MoSe$_2$ NSs and WS$_2$ NSs, while slight oxidation was found for WS$_2$ NSs [40]. Raman spectra of MoS$_2$, MoSe$_2$, and WS$_2$ NSs (Figure 6d) reveal the characteristic peaks for MoS$_2$ (408 cm^{-1} and 382 cm^{-1}), MoSe$_2$ (241 cm^{-1} and 287 cm^{-1}), and WS$_2$ (420 cm^{-1} and 354 cm^{-1}), which belong to their A$_{1g}$ and E$^1_{2g}$ vibrational modes, respectively [33]. Figure 6e shows the UV-vis-NIR spectra of the TMDC NSs aqueous dispersions and their photos under ambient light (inset). The distinct absorption peaks located near 800 nm (MoSe$_2$), 632 nm (MoS$_2$), and 620 nm (WS$_2$) originate from the A exciton transition, similar to the reports of few-layer TMDC NSs [43,50]. Figure 6f and Figure S11 show the FT-IR spectra of the TMDC NSs and their bulk materials. The IR absorption bands near 3410 cm^{-1} and 1630 cm^{-1} can be assigned to the stretching frequency of OH and COO groups from CMC adsorbed on the surface of the TMDC NSs.

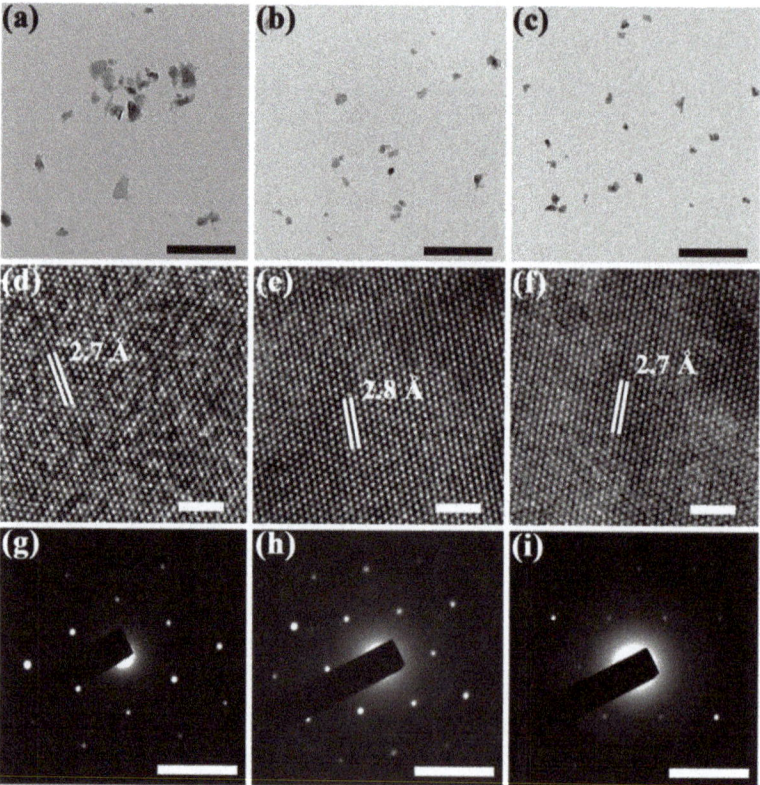

Figure 4. (a–c) TEM, (d–f) HRTEM, and (g–i) SAED images of MoS_2, $MoSe_2$, and WS_2 NSs prepared by CMC–assisted ball milling: MoS_2 (a,d,g), $MoSe_2$ (b,e,h), and WS_2 (e,f,i). Scale bars: 200 nm for (a–c), 2 nm for (d–f), and 5 1/nm for (g–i).

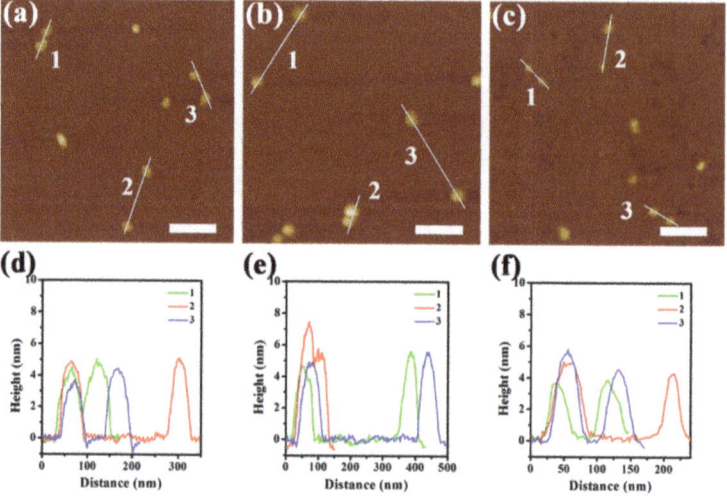

Figure 5. AFM images and corresponding height profiles of MoS_2, $MoSe_2$, and WS_2 NSs prepared by CMC–assisted ball milling: (a,d) MoS_2, (b,e) $MoSe_2$, (c,f) WS_2. Scale bar: 200 nm.

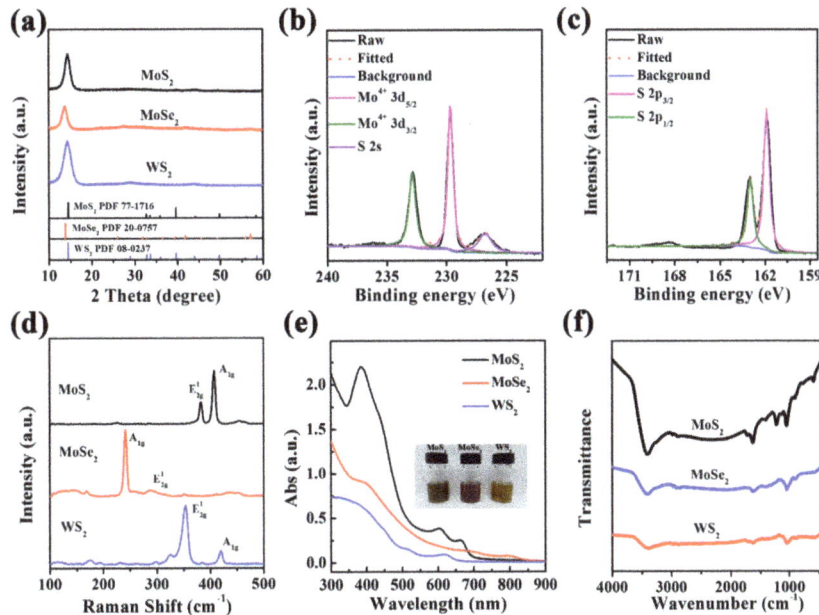

Figure 6. (**a**) XRD patterns of MoS$_2$, MoSe$_2$, and WS$_2$ NSs prepared by CMC–assisted ball milling. XPS spectra of MoS$_2$ NSs for Mo 3d (**b**) and S 2p (**c**) core level energy regions. (**d**) Raman spectra of MoS$_2$, MoSe$_2$, and WS$_2$ NSs. (**e**) UV–vis–NIR absorption spectra with the photos (inset) and (**f**) FT–IR spectra of MoS$_2$, MoSe$_2$, and WS$_2$ NSs.

3. Materials and Methods

3.1. Chemicals

WSe$_2$ powder (99.8%) was purchased from Alfa Aesar. MoS$_2$ powder (99%), MoSe$_2$ powder (99.9%), WS$_2$ powder (99%), Pluronic F-127, and polyvinyl pyrrolidone (PVP) were obtained from Sigma-Aldrich (St. Louis, MO, USA). Carboxymethyl cellulose sodium (CMC) and polyethylene glycol (PEG) were purchased from Aladdin and Macklin, respectively. Ultrapure water (18.2 MΩ, Billerica, MA, USA) was used to prepare aqueous solutions in this study.

3.2. Characterization

Transmission electron microscopy (TEM) images were obtained by using a HT7700 at 120 kV (Hitachi, Tokyo, Japan). High-resolution transmission electron microscopy (HRTEM) and selected area electron diffraction (SAED) images were obtained using field emission electron microscopes (Talos F200x, FEI, 200 kV, Waltham, MA, USA). Atomic force microscopy (AFM) images were acquired on Nanoscope IIIa (Bruker, Billerica, MA, USA). X-ray diffraction (XRD) patterns were recorded on an X-ray diffractometer (D8 Advance A25, Bruker, Billerica, MA, USA) with Cu Kα radiation (λ = 1.54178 Å). X-ray photoelectron spectroscopy (XPS) was performed on KRATOS Axis Supra (Shimadzu, Kyoto, Japan) with Al Kα (hν = 1486.6 eV) as the excitation source. Raman characterization was carried out on a micro-Raman spectroscopy system (inVia, Renishaw, Wotton-under-Edge, UK) equipped with a 532 nm laser. Ultraviolet-visible-near infrared (UV-vis-NIR) absorption spectroscopy was performed on a UV-3600 spectrophotometer (Shimadzu, Kyoto, Japan). Fourier transform infrared (FT-IR) spectra were recorded on a FT-IR spectrometer (Perkin Elmer, Waltham, MA, USA). The concentration of TMDC NSs was determined by using an inductively coupled plasma optical emission spectrometer (ICP-OES, Optima 5300DV, Perkin Elmer, Waltham, MA, USA).

3.3. Preparation of WSe$_2$ NSs by CMC-Assisted Ball Milling

WSe$_2$ powder (0.1 g) and CMC (1.0 g) were mixed and put into the steel jar of a planetary ball mill (Tianchuang, Changsha, China) for ball milling at 650 rpm (35× g) for 12 h. Then, 40 mL ultrapure water was added into the jar and mixed. The aqueous suspensions of WSe$_2$ were first centrifuged at 3000 rpm (765× g) for 30 min to remove the large aggregates. Then, the supernatant was centrifuged at 10,000 rpm (8497× g) for 1 h. The sediment was collected and redispersed in water, while the supernatant was further centrifuged at 16,000 rpm (21,752× g) for 1.5 h. Similarly, the sediment was collected and the supernatant was then centrifuged at 21,000 rpm (37,471× g) for 4 h. Each sediment was further purified more than three times by centrifugation to remove the redundant CMC. Finally, the sediments were dispersed in ultrapure water and stored at 4 °C. WSe$_2$ NSs obtained at different centrifugation speeds (low speed at 10,000 rpm (8497× g), medium speed at 16,000 rpm (21,752× g), and high speed at 21,000 rpm (37,471× g)) are referred to as WSe$_2$-Low, WSe$_2$-Medium, and WSe$_2$-High, respectively.

3.4. WSe$_2$ NSs Prepared by Using Different Polymers during Ball Milling

The preparation of WSe$_2$ NSs by using different polymers, including F127, PVP, and PEG, has similar experimental conditions to CMC-assisted ball milling. Since the stabilization abilities of different polymers are different, the centrifugation conditions used to purify WSe$_2$ NSs were adjusted according to the polymer used. For F127, WSe$_2$ NSs were obtained by gradient centrifugation at 5000 rpm (2124× g) for 20 min (WSe$_2$-Low), 7500 rpm (4779× g) for 20 min (WSe$_2$-Medium), and 10,000 rpm (8497× g) for 30 min (WSe$_2$-High). For PVP, WSe$_2$ NSs were obtained by gradient centrifugation at 10,000 rpm (8497× g) for 1 h (WSe$_2$-Low), 16,000 rpm (21,752× g) for 2 h (WSe$_2$-Medium), and 21,000 rpm (37471× g) for 4 h (WSe$_2$-High). For PEG, WSe$_2$ NSs were obtained by gradient centrifugation at 5000 rpm (2124× g) for 20 min (WSe$_2$-Low), 7500 rpm (4779× g) for 20 min (WSe$_2$-Medium), and 10,000 rpm (8497× g) for 30 min (WSe$_2$-High).

3.5. Preparation of Other TMDC NSs

MoS$_2$, MoSe$_2$, and WS$_2$ NSs were prepared using similar procedures to the CMC-assisted ball milling of WSe$_2$ NSs, except that the centrifugation conditions were adjusted according to the different materials.

4. Conclusions

In summary, we have developed a novel polymer-assisted ball milling method for the efficient preparation of TMDC NSs with small sizes. The as-prepared WSe$_2$ NSs by using CMC have small sizes (8–40 nm) with high yield (over 60%), which is not easy to achieve by using other top-down methods. The high efficiency of this method is attributed to the enhanced interaction of the 2D TMDCs and the milling balls because of the polymer used during ball milling. The size and thickness of the WSe$_2$ NSs can be adjusted by changing the rotation speed and milling time during ball milling and the centrifugation conditions during purification. Moreover, this polymer-assisted ball milling method can also be used to prepare other TMDC NSs, such as MoS$_2$, WS$_2$, and MoSe$_2$. The as-prepared TMDC NSs have good colloidal stability in PBS and cell culture medium. This study provides a highly efficient, simple, general, and scalable method for the preparation of TMDC NSs with small sizes, which may also be used for other 2D materials.

Supplementary Materials: The following supporting information can be downloaded at: https://www.mdpi.com/article/10.3390/molecules27227810/s1, Figure S1: Large-scale TEM images of WSe2 NSs prepared by CMC-assisted ball milling at 650 rpm for 12 h and centrifuged under different conditions; Figure S2: Large-scale AFM images of WSe2 NSs prepared by CMC-assisted ball milling at 650 rpm for 12 h and centrifuged under different conditions; Figure S3: XPS spectra of WSe2 NSs prepared by CMC-assisted ball milling at 650 rpm for 12 h; Figure S4: Photographs of CMC-WSe2 NSs dispersed in water, PBS, and DMEM for different times; Figure S5: TEM images of WSe2 NSs prepared by ball milling with different polymers under different centrifugal conditions; Figure S6: TEM images of WSe2 NSs prepared by CMC-assisted ball milling for different times after gradient centrifugation; Figure S7: TEM images of WSe2 NSs prepared by CMC-assisted ball milling with different rotation speeds after gradient centrifugation; Figure S8: Size statistics of (a) MoS2, (b) MoSe2, and (c) WS2 NSs prepared by CMC-assisted ball milling; Figure S9: Large-scale AFM images of (a) MoS2, (b) MoSe2, and (c) WS2 NSs prepared by CMC-assisted ball milling; Figure S10: High-resolution XPS spectra of Mo 3d (a) and Se 3d (b) core level energy regions for MoSe2, and W 4f (c) and S 2p (d) core level energy region for WS2; Figure S11: FT-IR spectra of bulk MoS2, MoSe2, and WS2.

Author Contributions: Conceptualization, L.Y. and D.Z.; investigation and methodology, Q.Z., F.X., D.Z., and P.L.; writing—original draft preparation, Q.Z. and L.Y.; writing—review and editing, L.W. and L.Y. All authors have read and agreed to the published version of the manuscript.

Funding: The authors are grateful to the Natural Science Foundation of Jiangsu Province (BK20191382), the Leading-edge Technology Programme of Jiangsu Natural Science Foundation (BK20212012), and the Open Research Fund of Jiangsu Key Laboratory for Biosensors (JKLB202204).

Institutional Review Board Statement: Not applicable.

Informed Consent Statement: Not applicable.

Data Availability Statement: The raw data will be available from the corresponding authors upon reasonable request.

Conflicts of Interest: The authors declare no conflict of interest.

Sample Availability: Samples of the TMDC NSs are available from the authors.

References

1. Chhowalla, M.; Shin, H.S.; Eda, G.; Li, L.-J.; Loh, K.P.; Zhang, H. The chemistry of two-dimensional layered transition metal dichalcogenide nanosheets. *Nat. Chem.* **2013**, *5*, 263–275. [PubMed]
2. Su, J.; Liu, G.; Liu, L.; Chen, J.; Hu, X.; Li, Y.; Li, H.; Zhai, T. Recent Advances in 2D Group VB Transition Metal Chalcogenides. *Small* **2021**, *17*, e2005411. [PubMed]
3. Manzeli, S.; Ovchinnikov, D.; Pasquier, D.; Yazyev, O.V.; Kis, A. 2D transition metal dichalcogenides. *Nat. Rev. Mater.* **2017**, *2*, 17033. [CrossRef]
4. Kim, Y.; Woo, W.J.; Kim, D.; Lee, S.; Chung, S.M.; Park, J.; Kim, H. Atomic-Layer-Deposition-Based 2D Transition Metal Chalcogenides: Synthesis, Modulation, and Applications. *Adv. Mater.* **2021**, *33*, e2005907. [CrossRef]
5. Huang, X.; Zeng, Z.; Zhang, H. Metal dichalcogenide nanosheets: Preparation, properties and applications. *Chem. Soc. Rev.* **2013**, *42*, 1934–1946.
6. Splendiani, A.; Sun, L.; Zhang, Y.B.; Li, T.S.; Kim, J.; Chim, C.Y.; Galli, G.; Wang, F. Emerging Photoluminescence in Monolayer MoS$_2$. *Nano. Lett.* **2010**, *10*, 1271–1275. [CrossRef] [PubMed]
7. Mak, K.F.; Lee, C.; Hone, J.; Shan, J.; Heinz, T.F. Atomically Thin MoS$_2$: A New Direct-Gap Semiconductor. *Phys. Rev. Lett.* **2010**, *105*, 136805. [CrossRef]
8. Radisavljevic, B.; Radenovic, A.; Brivio, J.; Giacometti, V.; Kis, A. Single-layer MoS2 transistors. *Nat. Nanotechnol.* **2011**, *6*, 147–150. [CrossRef] [PubMed]
9. Kibsgaard, J.; Chen, Z.; Reinecke, B.N.; Jaramillo, T.F. Engineering the surface structure of MoS$_2$ to preferentially expose active edge sites for electrocatalysis. *Nat. Mater.* **2012**, *11*, 963–969. [PubMed]
10. Liu, Y.; Li, J.; Yi, S.; Ge, X.; Chen, X.; Luo, J. Enhancement of friction performance of fluorinated graphene and molybdenum disulfide coating by microdimple arrays. *Carbon* **2020**, *167*, 122–131.
11. Liu, Y.; Chen, X.; Li, J.; Luo, J. Enhancement of friction performance enabled by a synergetic effect between graphene oxide and molybdenum disulfide. *Carbon* **2019**, *154*, 266–276.
12. Liu, Y.-F.; Liskiewicz, T.; Yerokhin, A.; Korenyi-Both, A.; Zabinski, J.; Lin, M.; Matthews, A.; Voevodin, A.A. Fretting wear behavior of duplex PEO/chameleon coating on Al alloy. *Surf. Coat. Technol.* **2018**, *352*, 238–246. [CrossRef]
13. Wang, X.; Sun, G.; Li, N.; Chen, P. Quantum dots derived from two-dimensional materials and their applications for catalysis and energy. *Chem. Soc. Rev.* **2016**, *45*, 2239–2262. [CrossRef] [PubMed]
14. Xu, Y.; Wang, X.; Zhang, W.L.; Lv, F.; Guo, S. Recent progress in two-dimensional inorganic quantum dots. *Chem. Soc. Rev.* **2018**, *47*, 586–625.
15. Li, X.; Shan, J.; Zhang, W.; Su, S.; Yuwen, L.; Wang, L. Recent Advances in Synthesis and Biomedical Applications of Two-Dimensional Transition Metal Dichalcogenide Nanosheets. *Small* **2017**, *13*, 1602660.
16. Han, J.H.; Kwak, M.; Kim, Y.; Cheon, J. Recent Advances in the Solution-Based Preparation of Two-Dimensional Layered Transition Metal Chalcogenide Nanostructures. *Chem. Rev.* **2018**, *118*, 6151–6188. [CrossRef] [PubMed]

17. Najmaei, S.; Liu, Z.; Zhou, W.; Zou, X.; Shi, G.; Lei, S.; Yakobson, B.I.; Idrobo, J.C.; Ajayan, P.M.; Lou, J. Vapour phase growth and grain boundary structure of molybdenum disulphide atomic layers. *Nat. Mater.* **2013**, *12*, 754–759. [PubMed]
18. Matte, H.; Gomathi, A.; Manna, A.K.; Late, D.J.; Datta, R.; Pati, S.K.; Rao, C.N.R. MoS$_2$ and WS$_2$ Analogues of Graphene. *Angew. Chem. Int. Ed.* **2010**, *49*, 4059–4062.
19. Hwang, H.; Kim, H.; Cho, J. MoS$_2$ Nanoplates Consisting of Disordered Graphene-like Layers for High Rate Lithium Battery Anode Materials. *Nano. Lett.* **2011**, *11*, 4826–4830. [CrossRef] [PubMed]
20. Novoselov, K.S.; Jiang, D.; Schedin, F.; Booth, T.J.; Khotkevich, V.V.; Morozov, S.V.; Geim, A.K. Two-dimensional atomic crystals. *Proc. Nat. Acad. Sci. USA* **2005**, *102*, 10451–10453.
21. Joensen, P.; Frindt, R.F.; Morrison, S.R. Single-layer MoS$_2$. *Mater. Res. Bull.* **1986**, *21*, 457–461.
22. Zeng, Z.; Yin, Z.; Huang, X.; Li, H.; He, Q.; Lu, G.; Boey, F.; Zhang, H. Single-Layer Semiconducting Nanosheets: High-Yield Preparation and Device Fabrication. *Angew. Chem. Int. Ed.* **2011**, *50*, 11093–11097.
23. Yuwen, L.; Yu, H.; Yang, X.; Zhou, J.; Zhang, Q.; Zhang, Y.; Luo, Z.; Su, S.; Wang, L. Rapid preparation of single-layer transition metal dichalcogenide nanosheets via ultrasonication enhanced lithium intercalation. *Chem. Commun.* **2016**, *52*, 529–532. [CrossRef] [PubMed]
24. Coleman, J.N.; Lotya, M.; O'Neill, A.; Bergin, S.D.; King, P.J.; Khan, U.; Young, K.; Gaucher, A.; De, S.; Smith, R.J.; et al. Two-Dimensional Nanosheets Produced by Liquid Exfoliation of Layered Materials. *Science* **2011**, *331*, 568–571. [CrossRef]
25. Smith, R.J.; King, P.J.; Lotya, M.; Wirtz, C.; Khan, U.; De, S.; O'Neill, A.; Duesberg, G.S.; Grunlan, J.C.; Moriarty, G.; et al. Large-Scale Exfoliation of Inorganic Layered Compounds in Aqueous Surfactant Solutions. *Adv. Mater.* **2011**, *23*, 3944–3948. [PubMed]
26. Zhao, X.; Ma, X.; Sun, J.; Li, D.; Yang, X. Enhanced Catalytic Activities of Surfactant-Assisted Exfoliated WS$_2$ Nanodots for Hydrogen Evolution. *ACS Nano* **2016**, *10*, 2159–2166. [PubMed]
27. Zhou, Y.; Xu, L.; Liu, M.; Qi, Z.; Wang, W.; Zhu, J.; Chen, S.; Yu, K.; Su, Y.; Ding, B.; et al. Viscous Solvent-Assisted Planetary Ball Milling for the Scalable Production of Large Ultrathin Two-Dimensional Materials. *ACS Nano* **2022**, *16*, 10179–10187. [PubMed]
28. Yao, Y.; Lin, Z.; Li, Z.; Song, X.; Moon, K.-S.; Wong, C.-P. Large-scale production of two-dimensional nanosheets. *J. Mater. Chem.* **2012**, *22*, 13494–13499.
29. Krishnamoorthy, K.; Pazhamalai, P.; Veerasubramani, G.K.; Kim, S.J. Mechanically delaminated few layered MoS$_2$ nanosheets based high performance wire type solid-state symmetric supercapacitors. *J. Power Sources* **2016**, *321*, 112–119.
30. Ashraf, W.; Khan, A.; Bansal, S.; Khanuja, M. Mechanical ball milling: A sustainable route to induce structural transformations in tungsten disulfide for its photocatalytic applications. *Phys. E Low-Dimens. Syst. Nanostructures* **2022**, *140*, 115152.
31. Yi, M.; Shen, Z. A review on mechanical exfoliation for the scalable production of graphene. *J. Mater. Chem. A* **2015**, *3*, 11700–11715.
32. Zhu, T.T.; Zhou, C.H.; Kabwe, F.B.; Wu, Q.Q.; Li, C.S.; Zhang, J.R. Exfoliation of montmorillonite and related properties of clay/polymer nanocomposites. *Appl. Clay Sci.* **2019**, *169*, 48–66.
33. Abdelkader, A.M.; Kinloch, I.A. Mechanochemical Exfoliation of 2D Crystals in Deep Eutectic Solvents. *ACS Sustain. Chem. Eng.* **2016**, *4*, 4465–4472.
34. Dong, H.; Chen, D.; Wang, K.; Zhang, R. High-Yield Preparation and Electrochemical Properties of Few-Layer MoS$_2$ Nanosheets by Exfoliating Natural Molybdenite Powders Directly via a Coupled Ultrasonication-Milling Process. *Nanoscale Res. Lett.* **2016**, *11*, 409. [CrossRef]
35. Heinze, T.; Koschella, A. Carboxymethyl Ethers of Cellulose and Starch—A Review. *Macromol. Symp.* **2005**, *223*, 13–40.
36. Jawaid, A.; Che, J.; Drummy, L.F.; Bultman, J.; Waite, A.; Hsiao, M.S.; Vaia, R.A. Redox Exfoliation of Layered Transition Metal Dichalcogenides. *ACS Nano* **2017**, *11*, 635–646. [PubMed]
37. Wang, Y.; Liu, Y.; Zhang, J.; Wu, J.; Xu, H.; Wen, X.; Zhang, X.; Tiwary, C.S.; Yang, W.; Vajtai, R.; et al. Cryo-mediated exfoliation and fracturing of layered materials into 2D quantum dots. *Sci. Adv.* **2017**, *3*, e1701500. [PubMed]
38. Yuwen, L.; Zhou, J.; Zhang, Y.; Zhang, Q.; Shan, J.; Luo, Z.; Weng, L.; Teng, Z.; Wang, L. Aqueous phase preparation of ultrasmall MoSe2 nanodots for efficient photothermal therapy of cancer cells. *Nanoscale* **2016**, *8*, 2720–2726. [PubMed]
39. Bang, G.S.; Cho, S.; Son, N.; Shim, G.W.; Cho, B.K.; Choi, S.Y. DNA-Assisted Exfoliation of Tungsten Dichalcogenides and Their Antibacterial Effect. *ACS Appl. Mater. Interfaces* **2016**, *8*, 1943–1950. [CrossRef] [PubMed]
40. Ambrosi, A.; Sofer, Z.; Pumera, M. 2H→1T phase transition and hydrogen evolution activity of MoS$_2$, MoSe$_2$, WS$_2$ and WSe$_2$ strongly depends on the MX$_2$ composition. *Chem. Commun.* **2015**, *51*, 8450–8453.
41. Wang, H.; Kong, D.; Johanes, P.; Cha, J.J.; Zheng, G.; Yan, K.; Liu, N.; Cui, Y. MoSe2 and WSe2 Nanofilms with Vertically Aligned Molecular Layers on Curved and Rough Surfaces. *Nano. Lett.* **2013**, *13*, 3426–3433.
42. Del Corro, E.; Terrones, H.; Elias, A.; Fantini, C.; Feng, S.; Nguyen, M.A.; Mallouk, T.E.; Terrones, M.; Pimenta, M.A. Excited excitonic states in 1L, 2L, 3L, and bulk WSe$_2$ observed by resonant Raman spectroscopy. *ACS Nano* **2014**, *8*, 9629–9635. [PubMed]
43. Zhao, W.; Ribeiro, R.M.; Eda, G. Electronic structure and optical signatures of semiconducting transition metal dichalcogenide nanosheets. *Acc. Chem. Res.* **2015**, *48*, 91–99. [PubMed]
44. Biswal, D.R.; Singh, R.P. Characterisation of carboxymethyl cellulose and polyacrylamide graft copolymer. *Carbohydr. Polym.* **2004**, *57*, 379–387.
45. He, F.; Zhao, D.; Liu, J.; Roberts, C.B. Stabilization of Fe–Pd Nanoparticles with Sodium Carboxymethyl Cellulose for Enhanced Transport and Dechlorination of Trichloroethylene in Soil and Groundwater. *Ind. Eng. Chem. Res.* **2007**, *46*, 29–34.

46. Guan, G.; Zhang, S.; Liu, S.; Cai, Y.; Low, M.; Teng, C.P.; Phang, I.Y.; Cheng, Y.; Duei, K.L.; Srinivasan, B.M.; et al. Protein Induces Layer-by-Layer Exfoliation of Transition Metal Dichalcogenides. *J. Am. Chem. Soc.* **2015**, *137*, 6152–6155. [PubMed]
47. Zong, L.; Li, M.; Li, C. Bioinspired Coupling of Inorganic Layered Nanomaterials with Marine Polysaccharides for Efficient Aqueous Exfoliation and Smart Actuating Hybrids. *Adv. Mater.* **2017**, *29*, 1604691. [CrossRef]
48. Kang, T.W.; Han, J.; Lee, S.; Hwang, I.J.; Jeon, S.J.; Ju, J.M.; Kim, M.J.; Yang, J.K.; Jun, B.; Lee, C.H.; et al. 2D transition metal dichalcogenides with glucan multivalency for antibody-free pathogen recognition. *Nat. Commun.* **2018**, *9*, 2549. [PubMed]
49. Eda, G.; Yamaguchi, H.; Voiry, D.; Fujita, T.; Chen, M.; Chhowalla, M. Photoluminescence from Chemically Exfoliated MoS_2. *Nano. Lett.* **2011**, *11*, 5111–5116. [PubMed]
50. Zhao, W.; Ghorannevis, Z.; Chu, L.; Toh, M.; Kloc, C.; Tan, P.-H.; Eda, G. Evolution of Electronic Structure in Atomically Thin Sheets of WS_2 and WSe_2. *ACS Nano* **2013**, *7*, 791–797.

Review

The Principle of Nanomaterials Based Surface Plasmon Resonance Biosensors and Its Potential for Dopamine Detection

Faten Bashar Kamal Eddin [1] and Yap Wing Fen [1,2,*]

1. Department of Physics, Faculty of Science, University Putra Malaysia, UPM, Serdang 43400, Selangor, Malaysia; faten.mphy@gmail.com
2. Functional Devices Laboratory, Institute of Advanced Technology, University Putra Malaysia, UPM, Serdang 43400, Selangor, Malaysia
* Correspondence: yapwingfen@gmail.com

Academic Editors: Minas M. Stylianakis, Athanasios Skouras and Ashok Kakkar
Received: 20 April 2020; Accepted: 25 May 2020; Published: 15 June 2020

Abstract: For a healthy life, the human biological system should work in order. Scheduled lifestyle and lack of nutrients usually lead to fluctuations in the biological entities levels such as neurotransmitters (NTs), proteins, and hormones, which in turns put the human health in risk. Dopamine (DA) is an extremely important catecholamine NT distributed in the central nervous system. Its level in the body controls the function of human metabolism, central nervous, renal, hormonal, and cardiovascular systems. It is closely related to the major domains of human cognition, feeling, and human desires, as well as learning. Several neurological disorders such as schizophrenia and Parkinson's disease are related to the extreme abnormalities in DA levels. Therefore, the development of an accurate, effective, and highly sensitive method for rapid determination of DA concentrations is desired. Up to now, different methods have been reported for DA detection such as electrochemical strategies, high-performance liquid chromatography, colorimetry, and capillary electrophoresis mass spectrometry. However, most of them have some limitations. Surface plasmon resonance (SPR) spectroscopy was widely used in biosensing. However, its use to detect NTs is still growing and has fascinated impressive attention of the scientific community. The focus in this concise review paper will be on the principle of SPR sensors and its operation mechanism, the factors that affect the sensor performance. The efficiency of SPR biosensors to detect several clinically related analytes will be mentioned. DA functions in the human body will be explained. Additionally, this review will cover the incorporation of nanomaterials into SPR biosensors and its potential for DA sensing with mention to its advantages and disadvantages.

Keywords: neurotransmitters; nanomaterials; surface plasmon resonance; optical; biosensors; diagnosis; dopamine

1. Introduction

Over the last few decades there has been a great effort towards the development of label-free optical biosensors. These essential analytical tools offer real-time analysis, detection of chemical and biological species with high sensitivity and selectivity. The tremendous advances in these biosensors will have a major impact on our health care. Among these technologies used to analyze the bio-specific interactions, surface plasmon resonance (SPR) biosensors today belong to the most advanced [1]. It has proven effective in medical diagnostics [2,3], food quality tests [4], detection of heavy metal ions [5], and others with respect to environmental protection.

Comparing to the conventional diagnostic tools, SPR biosensors have multiple advantages such as easy preparation, no requirement of labeling, real-time detection capability, cost- effectiveness, and high specificity and sensitivity. However, for the label-free detection of low concentrations of analytes with small molecular weight its sensitivity is not enough. Therefore, considerable efforts have been invested to overcome these challenges and improve the sensitivity of the SPR biosensor with keeping all its advantages. Nanomaterials are promising candidates and have demonstrated their appropriateness in the biosensing field. All nanomaterials have a general feature, which is the high specific surface. This enables the immobilization of an enhanced amount of bioreceptor units. Using the functional nanomaterials significantly enhanced the sensors performances, increased the sensitivity and selectivity of the sensing platform. The sensing performance is affected by synthetic procedure of the used nanomaterial, its shape and size. Additionally, the immobilization strategy used to functionalize the sensor chip is still challenge [6]. The purpose of this concise review is to introduce SPR concepts, and simplify the mechanism of SPR based sensor from dip to real-time measurements, explain the important characteristics in SPR sensor performance, mention several clinically related analytes that have been detected using SPR biosensors efficiently. Additionally, this mini review will explain the critical role of dopamine (DA) in human body and the potential of nanomaterials based SPR biosensors to detect it with mentioning its advantages and disadvantages.

2. SPR Phenomena

SPR is a quantum electromagnetic phenomenon that occurs when light interacts with free electrons at the interface between the metal and dielectric [7,8]. This optical process happens when monochromatic and p-polarized light beam strikes the surface of metal (typically gold) as shown in Figure 1.

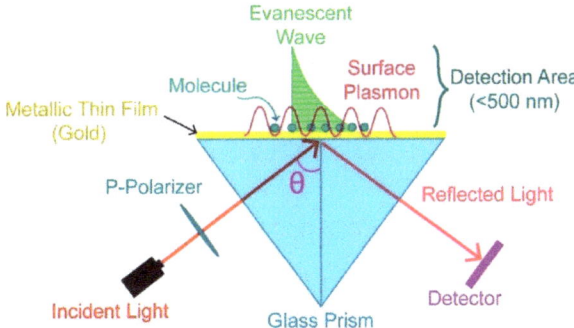

Figure 1. Experimental set-up of surface plasmons (SPs) excitation.

At a specific incidence angle when light satisfies resonance conditions and the frequency of the incident light matches the frequency of the surface plasmon wave, the light energy partially transfers to the electron packages on the metallic surface. After that, the observed reflected light shows a dip in the intensity as shown in Figure 2a. The electron coherent oscillations that were excited by exponentially decaying evanescent field of the incident light are called surface plasmons (SPs) and propagate parallel to the metallic surface. The angle at which the reflected light shows the maximum loss of intensity is called resonance angle or SPR-dip [9,10].

At the beginning of the twentieth century in 1902, the first observation of SPs was obtained by Wood, who reported anomalies in the diffraction spectrum of polychromatic light on a metallic diffraction grating [11]. Then, the connection between these abnormal diffractions and the excitation of electromagnetic surface waves on the diffraction grating surface was proved by Fano [12]. The clear explanation of this phenomenon was not complete until 1968, when Otto verified the concept experimentally and proved that in the attenuated total reflection (ATR) method, the excitation of SPs led to drop in the reflectivity [13]. Before the end of the same year, Kretschmann and Raether made some modifications on the configuration of ATR method to observe the excitation of SPs [14].

These important achievements done by Otto, Kretschmann, and Raether established an appropriate method to excite SPs and ushered in a promising future in modern optics. There are two categories of SPs, propagating SPs (PSPs), and localized SPs (LSPs). The excitation of SPs in the first category occurs on the metallic films. There are several approaches for this type of SPs such as the Kretschmann and Otto prism coupler, optical waveguides coupler, diffraction gratings, and optical fiber coupler. While the excitation of LSPs occurs on metallic nanoparticles [15]. The most common approach used in triggering SPs is prism coupling, which is also known as the ATR mode [16]. In Otto configuration, there is a certain distance separates the prism and the thin metallic film, the refractive index (RI) of this dielectric layer is small. The Kretschmann configuration is the easiest one. In this geometry, the prism is in contact with the metallic layer directly, and the wave vector component of SPs propagating along the interface is coupled with the wave vector of the incident photon. The practical performance of the sensor is confined to the resonance angle. The sensitivity of SPR sensors based on prisms is higher than that of SPR sensors based on grating couplers. Additionally, prism-based SPR sensors using wavelength interrogation provide the best resolution and a great deal of flexibility in terms of the analyte refractive index (RI) covered [17]. So, it has become a highly efficient mechanism for optical sensing of several biological, chemical and environmental changes.

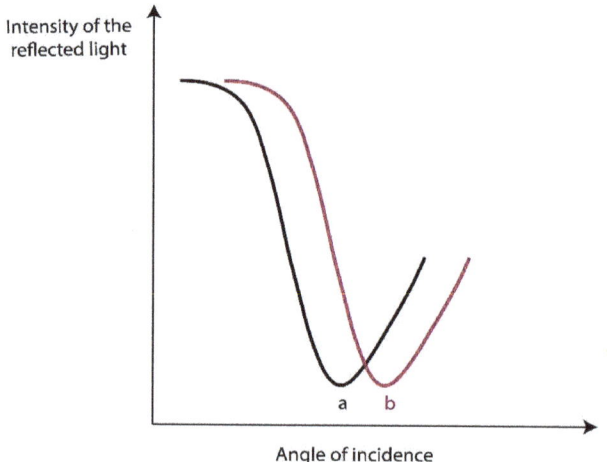

Figure 2. (a) A dip in the intensity of the reflected light after SPs excitation and (b) an angular shift from a to b due to a refractive index (RI) change on the Au film.

3. SPR Based Sensor

SPR is an excellent method to monitor changes occurring in the RI in the near vicinity of the metal surface. When the RI changes as a result of adsorption of molecules on the metal surface, the resonance spectral response of the SPR will change, and thus shifting angular or spectral position of the SPR dip will happen as shown in Figure 2b to reflect certain properties of the system and provide information on the kinetics of molecules adsorbed on the surface.

SPR sensors lack intrinsic selectivity, the change of the signal depends on all RI changes in the evanescent field. Changing the buffer composition or concentration leads to RI difference of the medium. Additionally, this depends on the medium temperature and the non-specific and specific adsorption of molecules on the SPR chips. The enhancement of the SPR biosensor needs modification of its surface with suitable ligands to capture the target compound (the analyte) and neglect other molecules available in the sample as shown in Figure 3. These ligands can be permanently or temporarily immobilized on the sensor surface. The analyte accumulation results in a RI change in the evanescent field detected. When the ligand captures the analyte, the measurable signal rises and this is called direct label free detection.

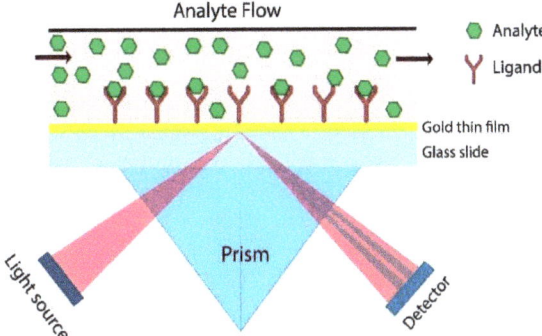

Figure 3. Direct label free detection.

Following in time the resonance angle or wave length shift at which the dip is observed produces the sensogram (Figure 4), then the amount of adsorbed species after injection of the original baseline buffer can be determined, and a study of the kinetics of the biomolecular interaction can be done. The sensor surface should be conditioned with a suitable buffer solution to start each measurement. It is essential to have a stable base line first [18]. After analyte injection, the association phase starts and the ligands on the surface of the sensor capture the target compounds. Upon injection of the baseline, the dissociation of target compounds and non-specifically binding molecules from the surface start. To break the specific binding between the analyte and ligands, regeneration solution should be injected. This step is vital in order to perform many tests using the same sensor surface.

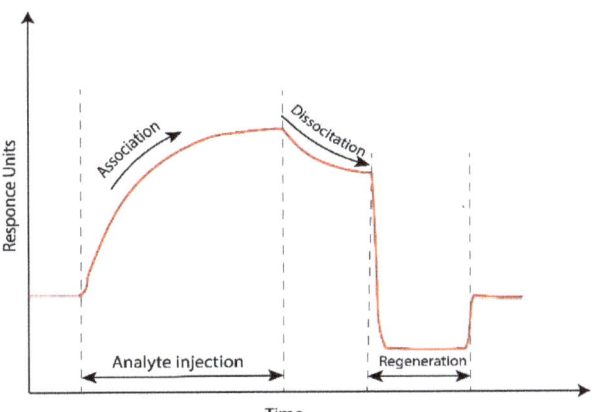

Figure 4. Surface plasmon resonance (SPR) sensogram.

4. The Important Characteristics in SPR Sensor Performance

To evaluate the performance of the SPR sensor, there are several major characteristics that should be taken in consideration. The main performance indicators include the sensor sensitivity, resolution, accuracy, reproducibility, repeatability, and limit of detection (LOD) [19]. The ratio of the change in SPR sensor output (e.g., angle of incidence, wavelength, amplitude) to the change in the measurand (e.g., RI, analyte concentration, and thickness) represents the sensitivity of the sensor. The smallest change in the RI that produces a detectable change in the SPR sensor output defines the sensor resolution.

The accuracy defines the degree to which the sensor readout value corresponds to the actual value of the measurand. Sensor's precision also includes reproducibility, which shows whether entire measurements can be reproduced in its entirety. Additionally, the repeatability is a way to measure the sensor precision. It shows the sensor ability to reproduce the same response under the same conditions over many repetitions. The LOD means the lowest concentration of analyte that can be detected by the sensor.

The shape of the SPR response curve affects the two important performance parameters, sensitivity and signal-to-noise ratio (SNR). Sharp SPR-dip is desirable because the narrower the width, the higher the detection accuracy and the deeper the curve, the higher the sensitivity. The sensor design parameters such as light polarization modes, light coupling techniques, types of metals influence the width and depth of the SPR dip. In prism-based SPR sensor, the physical structure of prism used (triangular, conical, hemispherical, and half cylindrical) and its RI affects the SPR curve [20]. Among the parameters that play a prominent role in the development of a highly sensitive SPR sensor, the thickness of the metallic thin films must be mentioned [21]. Until now, the ideal thickness of prepared layers to generate maximum SPR is within the range of 40 nm to 60 nm [22,23]. Additionally, the used wavelengths for excitation strongly affect the resonance curve width. Narrowing the reflectance curve necessarily increases the SPR propagation length [24,25]. Using the excitation wavelength in the IR region results in an increase in the penetration depth with the consequence that the reflectance minimum will become more sensitive to dielectric changes relatively far from the metal/dielectric interface; thus, the SPR signal gets weaker and the surface-sensitive character of SPR becomes less prominent. The usage of red laser, λ = 633 nm significantly enhances the maximum excitation of SPs and leads to strong SPR signal, while the excitation wavelength in UV region generates a very weak SPR signal [26].

5. The Importance of SPR Biosensors in the Medical Diagnosis

Many years ago, SPR technology firstly emerged and then many scientists in various fields include chemistry, biology, physics, and medicine have joined to use this promising technology. Recently, SPR biosensors have fascinated impressive attention as medical diagnostic tools due to many reasons. It is easy to prepare them with low cost and without labeling. These sensors provide high specificity and sensitivity, and are capable to detect clinically relevant analyte in real-time. SPR as a simple and direct sensing approach has been used to detect different clinical entities. SPR sensor was used to diagnostic human hepatitis B virus antibodies [27,28]. Additionally, SPR sensor platforms were developed for total prostate-specific antigen detection [29,30]. The extraordinary properties of graphene were exploited to construct SPR sensor to detect folic acid protein [31]. Magnetic nanoparticles with core-shell structure added amplification technique to the SPR sensor to detect α-fetoprotein [32]. The combination of SPR sensor with advantages of molecular imprinting-based synthetic receptors was reported in the detection of cardiac biomarkers used to diagnose acute myocardial infarction such as myoglobin, creatine kinase-myocardial band, and cardiac troponins [33]. SPR chips were modified using lysozyme imprinted polymeric nanoparticles to detect the changes in lysozyme levels, which work as indicators for some diseases including leukemia, meningitis, several kidney problems, and rheumatoid arthritis [34]. It was demonstrated that SPR sensor has the ability to detect pregnancy associated plasma protein A2, which is a metalloproteinase that plays multiple roles in fetal development and postnatal growth [35]. Several studies were reported on using SPR sensor as a new strategy to detect influenza nucleoprotein [36], avian influenza A H7N9 virus [37], maize chlorotic mottle virus [38], herpes simplex virus (HSV) [39], cancer cell line (HeLa cells) with biomarker Rhodamine 6G related to cancer tumors [40], nonstructural protein 1 of Zika virus [41], non-human pathogen feline calicivirus (FCV) [42], and dengue virus (DENV) [43–46].

SPR technology proved its efficiency again in clinical field as a diagnostic tool of the endocrine diseases. Using the SPR sensor, hormone levels can be directly monitored and measured. By immobilization of the molecularly imprinted nanoparticles onto the SPR chip it was easy to detect iron regulating hormone Hepcidin-25 [47]. Other modified SPR sensors were employed to detect testosterone [48,49], gonadotropic hormones and luteinizing hormone [50], pituitary hormones such as human thyroid-stimulating, growth, follicle-stimulating [51], and insulin [52]. There are other reviews and articles that focus on the importance and the application of SPR biosensors for the diagnosis of medically important entities such as viruses, neurotransmitters, proteins, hormones, nucleic acids, cells, drugs, and disease biomarkers [53–57].

6. DA and Its Critical Role in the Human Body

In mammalian central nervous systems, during the synaptic transmission process neurons secrete endogenous chemical messengers that are called neurotransmitters (NTs). NTs transmit signals across synapses from one neuron cell to a target neuron cell as shown in Figure 5 or to muscle cells, gland cells and other non-neuronal body cells. Thus, NTs relay information throughout the brain and the whole-body. DA is one of the most crucially important NTs, it plays a vital role in the neural functions like information flow and attention span, consciousness, learning, motions, emotions and memory formation. In addition to its critical roles related to renal, hormonal and cardiovascular systems [58–61].

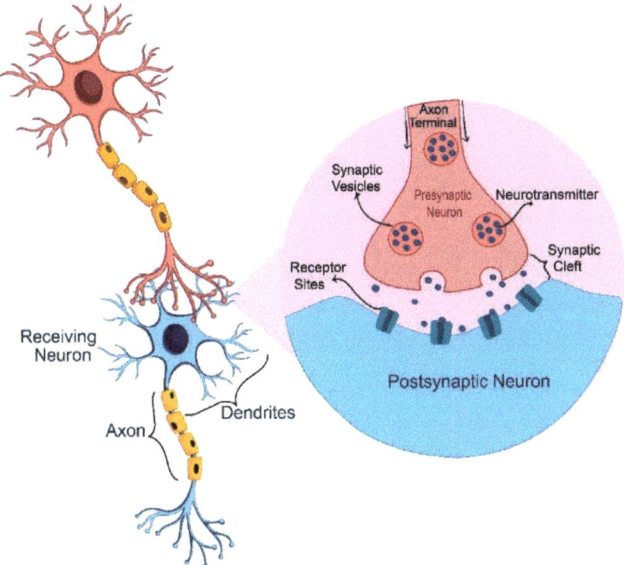

Figure 5. Neuron communication process.

Abnormal concentrations of DA in different biological fluids are associated with various diseases. Cardiotoxicity and subsequent rapid heart rate, hypertension, and heart failure can be an indicator of the high levels of DA [62]. While deficiency or practically complete depletion of DA may result in various neurodegenerative diseases such as Parkinson's disease (PD) [63,64], Alzheimer's disease [65,66], schizophrenia [67–75], and depression [76]. Therefore, the fabrication and development of highly sensitive and selective sensors for quantitative determination of DA in vivo and in vitro is extremely necessary in the clinical diagnosis, it can make great contribution monitoring the effectiveness of the treatment, prevention of diseases [77]. The monitoring of DA levels can be done in different biological samples including saliva, urine, plasma, serum, platelets, and cerebral spinal fluid [78,79]. DA physiological levels in humans vary in these biofluids. According to the Human Metabolome Database, DA concentration is less than 130 pM in blood, while its levels in human urine and cerebrospinal fluid are approximately 5 nM [80]. To date, the development of analytical methods to measure the concentration of DA directly continues to grow. A variety of these methods demonstrated its capability to detect low levels of DA including high performance liquid chromatography (HPLC) [81–83], capillary electrophoresis [84–87], Fourier transform infrared (FTIR) spectroscopy [88], flow injection [89], enzymatic methods [90], electrochemical (EC) methods [91–93], mass spectroscopy [94–96], and various types of optical methods such as colorimetry and spectrophotometry [97], fluorescence [98–101], electrochemiluminescence (ECL) [102], surface-enhanced Raman spectroscopy (SERS) [103–108], chemiluminescence (CL) [109], photoelectrochemical (PEC), photoluminescence (PL), solid phase spectrophotometry (SPS), resonance Rayleigh scattering (RRS), and SPR spectroscopy [110–114]. The interested reader in electrochemical and optical methods is referred to the review paper by Kamal Eddin and Fen (2020) [115], and references cited therein.

7. DA Detection Using SPR Biosensors

Direct detection using SPR biosensor has the most notable benefit, which is the determination of the kinetics of biomolecular interactions. To study biomolecular interactions using SPR, there is no need to understand the optical phenomena in all its details. It is enough to know that SPR based sensors use an optical method to measure the RI changes near a sensor surface (within 300 nm from the surface). The SPR based sensor does not need complex equipment. During the design of this sensor, the optical unit, the liquid-containing unit and the sensor surface are combined into one system. The main focus in the preceding sections in this review is on SPR physical basics. However, it is necessary also to focus on the sensor chip and its surface chemistry where the biomolecular interactions take place. The few nanometers thickness of the coating and its morphology affects the SPR sensor performance and the quality of the obtained data. The SPR optical sensor had been used to characterize biomolecular interactions qualitatively and quantitatively due to their unparalleled advantages. The unique properties of SPR sensor made it a versatile tool used in various applications such as health care and medical diagnostics [116,117], food control [118,119], environmental pollutant monitoring [120–126], and others. However, their use to detect NTs is still growing and has attracted the attention of the scientific community.

SPR sensors detect the change in dielectric constants near noble metal film on the sensor chips. Therefore, the concern during the sensor design is on the surface modification and the immobilization of molecular recognition elements such as proteins, DNA, natural, and synthetic polymers to increase the selectivity to the detected analyte. The employment of molecularly imprinted polymer (MIP) with immobilized Au nanoparticles (NPs) (Au-MIP) to SPR sensors as a recognition element during DA sensing was reported by Matsui et al. (2005) [127]. Sensor chip modification was conducted in two steps, firstly the Au substrate was modified with MIP without Au NPs to avoid immobilization of Au NPs in too close vicinity to the Au substrate, which in turn reduces the sensitivity, then it was modified with Au NPs-MIP. The MIP swelling by incorporating water in accordance with analyte binding changes the dielectric constant near the surface of Au substrate significantly. More importantly, the distance between the Au NPs within the polymer gel and Au film on the sensor chip surface would be increased because of this swelling, which enhances the degree of SPR angle shift strongly. The modified sensor chip showed an increasing SPR angle in response to DA concentration in the range (1 nM to 1 mM). Furthermore, the Au NPs effectively enhanced the signal intensity (the change of SPR angle) in comparison with a sensor chip without Au NPs. Temperature-dependent behaviors of the sensor chips was also investigated by measuring SPR angle at different temperature. Au-MIP exhibited swelling in response to the low temperature. The proposed sensor demonstrated its repeatability, this because the analyte binding process and the consequent swelling was reversible.

The sensitivity of the Au-MIP-based SPR sensor can be also emphasized by comparison with the colorimetric sensor using a spectrophotometer [128], in which Au-MIP exhibited significant spectroscopic changes at 5 μM or higher concentrations.

The utilization of natural receptor to develop SPR biosensor was reported by Kumbhat et al. (2007) [129]. They developed an SPR based affinity biosensor using a D_3-DA receptor (DA-RC) as a recognition molecule for DA detection. During the immobilization of DA with bovine serum albumin

(BSA) protein (DA–BSA) conjugate on the sensor chip by physical adsorption, the loss of activity of biomolecules was minimized. The DA–BSA conjugate flowed over the gold surface, the increase in the resonance angle was an indicator on the saturation of the immobilization onto the sensor surface. Different concentrations (5–400 μg/mL) of the conjugate were used in the immobilization. The SPR angle shift reached a plateau around 100 μg/mL of the DA–BSA conjugate. The principle of indirect competitive inhibition to detect DA was employed in this work. DA-RC was allowed to flow over the DA–BSA surface, the resonance angle increased due to the binding of the receptor to the DA haptens present on the sensor surface. After completion of the DA-RC flow, the flow was switched back to phosphate buffered saline (PBS) buffer and the resonance angle shift remained almost stable due to the strong affinity interaction between DA-RC and DA–BSA conjugate. The sensor sensograms were observed for the affinity reaction between immobilized DA–BSA conjugate and DA-RC in the absence of DA, and in the presence of different concentrations of DA. The shift in the resonance angle decreased by increasing in the concentration of free DA in solution, this was because the free DA inhibited the binding interaction of DA-RC with the DA–BSA conjugate. The sensor exhibited a linear detection range from 0.085 to 700 ng/mL with a lower LOD of 0.085 ng/mL. The results showed that there is no significant interference species such as ascorbic acid (AA), uric acid (UA), and other DA analogues viz., DOPAC and 3-(3,4 dihydroxyphenyl)-alanine (DOPA). The stability of the sensor surface was high during repeated regeneration and affinity reaction cycles. According to the simplicity and effectiveness of this biosensor, their study presented an encouraging scope to develop portable detection systems to measure DA in vitro and in vivo in clinical and medical diagnostics.

The adsorption capability of Ibuprofen and DA on pure gold layer and gold functionalized with L-cysteine (L-Cys) and L-glutathione (L-GSH) using SPR spectroscopy was investigated by Sebők et al. (2013) [112]. From the description of DA adsorption by a two-stage isotherm it could be concluded that first layer of DA was irreversibly bound while the second layer was anchored by physical forces, so DA adsorption on the gold surface was only partially reversible. The amino groups available in DA enabled formation of a stronger chemical bond with the cysteine molecule comparing with ibuprofen. The bounded DA on the gold surface functionalized with glutathione showed perpendicular orientation to the gold surface, while glutathione orientation was the same as ibuprofen parallel to the surface. The initial energies of glutathione/ibuprofen interaction were lower comparing with the glutathione/DA system.

Among two-dimensional materials used to design high performance sensor platform, graphene has received profound interest due to its remarkable properties. In the work reported by Kamali et al. (2015), they developed the silver @ graphene oxide (Ag@GO) nanocomposite-based SPR optical sensor to detect DA and other biomolecules such as AA and UA [130]. The nanocomposite showed an SPR band at 402 nm due to the Ag NPs. The SPR intensity-based LOD of DA was 49 nM in the DA concentration linear range (100 nM to 1 μM), and 62 nM in the range (1–2 μM). While the SPR band position-based LOD of DA was 30 nM. The SPR absorbance peak changes toward UA and AA were not comparable with DA, which proved its excellent selectivity. The adsorption and sensing ability of the Ag@GO nanocomposite mainly depended on the nature of the adsorption site and its interaction with the functional groups of the molecules. The results proved that DA had more affinity with Ag@GO than UA and DA.

Next, Rithesh Raj et al. (2016) used green synthesized Ag NPs as sensing material to fabricate SPR based fiber optic sensor for DA measurement [131]. SPR spectra showed decrease in SPR dip intensity. Additionally, a shift in resonance wavelength towards lower wavelength was observed by increasing the concentrations of DA, this probably occurred because of the interaction between DA and Ag clusters which led to formation of Ag NPs. Poly vinyl alcohol (PVA) coated fiber showed negligible shift in wavelength comparing to green Ag NPs coated fiber, which demonstrated the role of the green Ag NPs in the sensing enhancement. The sensor response time was 6 min and the corresponding maximum resonance wavelength shift was 12 nm. The LOD was 2×10^{-7} M enhanced the selectivity of the proposed sensor in the presence of other biological species.

The utilization of conjugated polymer layer consisted of P(NIPAAm-st-MAAmBO) and glycopolymer (poly (2-lactobionamidoethyl methacrylamide)(PLAEMA) to functionalize the SPR sensing platform for DA detection was reported by Jiang et al. (2017) [132]. The SPR sensing mechanism was based on DA induced swelling of the conjugated polymer layer thickness, which increased the local IR and the reflection angle. This reversible swelling allows regeneration of the sensor. The increase in the reflection angle was measured as the sensogram change for quantitative analysis. By changing DA concentration in the range 1–10 nM, the sensogram exhibited a linear relation with DA concentration with a LOD of 1 nM. Increasing DA concentration led to an increase in the rate of adsorption because the multiple binding sites become more available on the P(NIPAAm$_{149}$-st-MAAmBO$_{19}$) chain. This in turn reduces the time of detection. By injection higher concentrations of DA, the binding sites were saturated then the resolution of the conjugated polymer functionalized sensor decreased. The proposed sensor had a broad dynamic range of 10^{-9} to 10^{-4} M.

The selectivity of the sensor was tested by comparing sensogram variation caused by different injections with 5 nM DA, glucose, AA and UA samples using SPR sensors functionalized with conjugated P(NIPAAm$_{149}$-st-MAAmBO$_{19}$) and P(LAEMA$_{21}$); P(NIPAAm$_{149}$-st-MAAmBO$_{19}$) only. The conjugated polymer functionalized sensor was selective to DA only. Additionally, the developed sensor was able to recognizing DA with coexisting saccharides. Polymer functionalized sensor provided much improved stability for at least 2 weeks in comparison with methods using specific DA receptors [133].

In the same year Abd Manaf et al. (2017) presented novel SPR sensor to detect DA down to 50 pM with high sensitivity of 36 nm/nM [134]. They developed a four-layer coating structure including SU8 waveguide, platinum, platinum NPs, and plastic for highly sensitive and selective measurements. Using Pt NPs, the effective surface area efficiently increased and the interaction between coating and waveguide channel increased in comparison with the metal coating without modification. The proposed sensor was designed for operation at high wavelength ranging between 1450 and 1600 nm to get high compatibility with normal optic fibers and achieve low power loss. The partial absorption of the input signal occurred through DA and Pt, and a power loss was noted at the output depending on the refractive indices of Pt and DA. Depending on the properties of the material used, in a certain wavelength, the absorption was at maximum (SPR dip). The RI of the DA solution was measured in order to identify the optimal thickness of the Pt layer using a prism coupler at 1550 nm wavelength. At a thickness below 50 nm of the Pt layer, too low sensitivity was obtained. While no significant improvement was observed when the thickness was more than 50 nm. Therefore, the suitable Pt layer was with a thickness of 50 nm. With changing DA concentration from 1 to 10 nM. When DA concentration decreases, SPR spectra shows shift in the resonance wavelength occurs toward a shorter wavelength. The dip was observed in the range 1500 and 1530 nm of wavelength. The sensor selectivity was high in the presence of glucose, lysine, and AA. The power loss was not significant at long wavelength, which qualifies this sensor to be used in high-precision application.

Recently, Cao and McDermott (2018) reported an ultrasensitive and selective method for DA detection. They incorporate DA DNA aptamer (DAAPT) conjugated AuNP that enhanced the inhibition assay using SPR imaging to detect and quantitate DA down to 200 fM. To the best of our knowledge, this is the lowest LOD achieved for SPR sensing of DA until now [80]. All the spectra have a good shape with the LSPR peaks located around 520 nm. The successful conjugation of DAAPT to AuNP surface was confirmed by observed the red shifts of the LSPR peaks. By mixing the 10 nm DAAPT-AuNP conjugate solution with DA solution at specific concentration, the binding between DA molecules and DAAPT-AuNP conjugates occurred, which in turns decreased the effective concentration of DAAPT-AuNP conjugate that can bind to the complementary DNA (cDNA) probe immobilized on a gold chip. With increasing DA concentrations, all the DAAPTs on the AuNP conjugate surface had the chance to bind to their DA targets so the conjugate probe is in its "OFF" state. In this case, the DNA aptamer was blocked and could not bind to cDNA on the chip surface, thus no signal response was observed. On the contrary, in the absence of DA, the binding between DAAPT on AuNP surface and DA did not happen, then the conjugate probe was in its "ON" state. Additionally, the binding of the conjugate to the cDNA probe was detected strongly. As a result, a big signal response was generated. In this study, the target DA molecule was acting as a switch to turn "OFF" and "ON" the DAAPT-AuNP conjugate during the whole process. The proposed assay showed good sensitivity, reproducibility, and its specificity to DA was high. Using this nanoparticle enhanced SPR aptasensor has the potential to detect small molecules, proteins, and other analytes, as long as the target has a corresponding aptamer.

Additionally, in 2018, Sharma used molecular imprinted graphene nanoplatelets (GNP)/tin oxide (SnO_2) nanocomposite to fabricate a fiber optic SPR-based DA sensor with high selectivity. The LOD of this sensor was lower than the LOD values of other DA sensors, it was 0.031 µM [135]. The dipping time of sensing element over the silver coated probe greater than 20 min produces sensing layer with a large thickness, which perturbs the intensity of the electric field at the interface and decreases the shift. By increasing DA concentration, the blue shift with a decrease in the depth of the SPR spectra was recorded. The change in the real and imaginary parts of the RI of the sensing layer occurred as a result of the binding of DA molecules with active sites in the sensing layer. The decrease in the real part of the effective RI led to that observed blue shift.

Recently, Sun et al. (2019) proposed an Au film/graphene D-shaped plastic optical fiber (D-POF) functionalized with DA binding aptamer (DBA) to detect DA sensitively and selectively in the presence of three different interferences. They demonstrated that graphene enhanced the sensitivity and played a role in amplification of the SPR signal by introducing DBA. They demonstrated the potential of this sensor to detect DA-induced disorders [136]. Graphene improved the sensor sensitivity and amplified the SPR signal. As RI changed from 1.3330 to 1.3612, the resonant wavelength of the proposed sensor had a red shift of 43.4 nm, which was larger than that of the sensor covered only by Au film 34.9 nm. Adding graphene increased the sensor sensitivity from 1238 to 1539 nm/RIU. The SPR dips decreased as the RI increased due to the higher loss of energy and the increasing in the penetration depth of evanescent field.

In the same year, Yuan et al. used DA to modify a gold sensing surface and produce self-assembled monolayers for secondary surface mediated reactions under different environmental conditions including pH, different buffers, varying DA concentrations in buffer solutions as well as the immobilization time of DA [137]. The favorable environmental conditions for DA immobilization onto gold surface were only alkaline (pH 7.6) and mildly alkaline media (pH 8.6 or 9.5) for 1 h. The immobilization for 2 h appeared to occur only in pH 7.6 or pH 8.6, and the SPR phenomenon did not exist at pH 9.5. This is because the thickness of the immobilized DA was over a critical reflectivity value. Using 2 mg/mL DA in pH 8.6 Tris buffer resulted in the optimum reactive on the gold chip. The critical DA immobilization time for SPR disappearance under the previous favorable conditions was estimated to be 277 min. These results provide valuable information for using DA for surface modification in the future studies. The recent works that used SPR phenomenon in DA measurements are summarized and discussed in Table 1.

Table 1. SPR sensors for dopamine (DA) detection.

Material	LOD	Detection Range	References
MIP-Au electrode	1 pM	-	[110]
DA-RC	0.085 ng/mL	0.085 ng/mL–700 ng/mL	[129]
DA antibodies/Au NPs/ITO	1 nM	0.001–100 µM	[113]
Ag@GO	30 nM	100 nM–2 µM	[130]
Ag NPs	0.2 µM	0.2–30 µM	[131]
Conjugated polymer P(NIPAAm149-st-MAAmBO19) and P(LAEMA21)	1 nM	1 nM–0.1 mM	[132]
Pt	50 pM	0.1 nM–32 µM	[134]
DAAPT-AuNPs	200 fM	100 µM–2 mM 200 fM–20 nM	[80]
Molecular Imprinted GNP/SnO_2 Nanocomposite	31 nM	0–100 µM	[135]
Au/graphene/DBA D-POF	-	0.1 nM–1 µM	[136]

8. The Advantages of DA Detection Using SPR Sensors

To evaluate the performance of any sensor, it is necessary to focus on some parameters that reveal its validity and efficiency. Among the most important of these parameters are the sensitivity and LOD, which must be sufficient enough for the low concentrations of the target. Additionally, the selectivity is very important to distinguish the target in the real sample where different interfering species exist. Reproducibility and overall reliability are also required. The detection limits of most sensors developed to detect DA were in the micromolar level. However, these sensors are not suitable for clinical diagnostics, which requires a nanomolar level of detection. Over the years, practical and economical SPR sensors have proven their efficiency in several fields as previously mentioned due to their overwhelming advantages. By employing SPR sensors, the reported LODs for DA are typically in the pM range or higher (nM). Recently, the LOD value in the range fM was obtained using this sensor. Table 2 shows a comparison between the lowest detection limit of different DA sensors. Obviously, the lowest LODs were in fM range obtained using EC, SERS, and SPR sensors. Despite the promising LODs obtained using EC and SERS sensors, they still have several disadvantages.

Table 2. Comparison of the detection limits of various DA sensors.

Method	Lowest Detection Limit	References
EC	78 fM	[138]
CL	0.19 nM	[139]
ECL	0.31 pM	[140]
Fluorescence	0.1 pM	[141]
Spectrophotometry	0.4 nM	[142]
Colorimetry	0.16 nM	[143]
SERS	0.006 pM	[144]
RRS	0.392 ng/mL	[145]
PRRS	0.1 pM	[146]
SPS	1.7 µM	[147]
PL	10 nM	[148]
Absorption	1.2 nM	[149]
PEC	2.3 pM	[150]
SPR	200 fM	[80]

E—Electrochemical; CL—Chemiluminescence; ECL—Electrochemiluminescence; SERS—Surface-enhanced Raman spectroscopy; RRS—Resonance Rayleigh scattering; PRRS—Plasmonic resonance Rayleigh scattering; SPS—Solid phase spectrophotometry; PL—Photoluminescence; PEC—Photoelectrochemical; SPR—Surface plasmon resonance.

An EC sensor offers real-time detection with simplicity and excellent sensitivity, fast response time, wide linear concentration range, cost effectiveness, and an ability to be miniaturized. However, it suffers from limited selectivity in the presence of other biological analytes, large noise and background signal. In addition to the fouling of the sensor surface and its degradation over time. Although the sensitivity and selectivity of the SERS-based DA sensors are higher in comparison to other detection methods. However, there is an obstacle to the availability of these sensors, they require expensive equipment for analysis. An SPR sensor has several important advantages such as direct label free detection of diverse molecule sets including small molecules, real-time measurements, high reliability, very high sensitivity with low detection limit, long-term stability, cost-effectiveness, easy sample preparation, small consumption of sample and reagent, reproducibility, regeneration of sensor chips, and specificity to the binding event. On the other hand, the disadvantages of this sensor are non-specific binding to surfaces needs to be controlled, which requires a meticulous experimental design, mass transport limitations when the analyte transfer to the ligand is limited, the immobilization effects, and the ligand can change its orientation after immobilization on the sensor chip and prevents binding with the analyte [151]. The combination of SERS with SPR has the potential to massively increase the local electromagnetic field intensity of nanoparticles to a level far exceeding the single-molecule SERS detection limit but this still lacks versatility [152]. However, in the context of a point-of-care testing (POCT) configuration, the bulky nature of the SPR apparatus hinders its field applications. So, the efforts are mainly made on the miniaturization of SPR devices.

9. Conclusions

In recent decades, the development of biosensors to identify and measure low concentrations of different NTs with high selectivity, sensitivity, low-cost, and rapid response attracted considerable attention due to the crucial role that neurotransmitters play in clinical diagnostics and curing mental disorders such as schizophrenia, Parkinson's disease, and Alzheimer's disease. SPR has emerged as a very suitable technology for clinical analytes detection. It is based on IR variations due to the mass change at the sensor chip. The SPR sensing platform has many features. It is easy to prepare it, the basic optics that it needs can be miniaturized to a suitable size in diagnosis, the direct quantification of the specific bindings occurs with high specificity and sensitivity. This low-cost technology is not based on labeling of the target molecules and does not alter its binding affinity and kinetics properties. Rapid and reliable detection of various medically important entities using SPR spectroscopy has been done over the last years. The reported works on DA sensing using this promising method is still limited despite the impressive results obtained. This is a strong motivation for researchers to develop these sensors by modification SPR chips using nanomaterials. Ongoing research employing the unique capabilities of carbon-based nanomaterials as well as the advantages of polymers and incorporating them with SPR technology for ultra-sensitive and selective detection of DA may introduce exciting progress in neuroscience. The high potential of SPR sensors and the continuous efforts to develop them and overcome their limitations qualify them to have a prominent presence in future developments in lab-on-a-chip technologies including point-of-care devices.

Author Contributions: Conceptualization, Y.W.F. and F.B.K.E.; Original Draft Preparation, F.B.K.E.; Writing and Editing, F.B.K.E.; Review, Y.W.F.; Visualization, F.B.K.E.; Software, F.B.K.E.; Resources, F.B.K.E.; Validation, Y.W.F.; Supervision, Y.W.F. All authors have read and agreed to the published version of the manuscript.

Funding: This research was funded by the Ministry of Education Malaysia through the Fundamental Research Grant Scheme (FRGS/1/2019/STG02/UPM/02/1).

Acknowledgments: F.B.K.E. acknowledges the support received from OWSD and Sida (Swedish International Development Cooperation Agency).

Conflicts of Interest: The authors declare no conflict of interest.

References

1. Kan, X.; Li, S.F.Y. Rapid Detection of Bacteria in Food by Surface Plasmon Resonance Sensors. *Int. J. Adv. Sci. Eng. Technol.* **2016**, *1*, 26–29.
2. Yanase, Y.; Hiragun, T.; Ishii, K.; Kawaguchi, T.; Yanase, T.; Kawai, M.; Sakamoto, K.; Hide, M. Surface Plasmon Resonance for Cell-Based Clinical Diagnosis. *Sensors* **2014**, *14*, 4948–4959. [CrossRef] [PubMed]
3. Yanase, Y.; Sakamoto, K.; Kobayashi, K.; Hide, M. Diagnosis of Immediate-Type Allergy Using Surface Plasmon Resonance. *Opt. Mater. Express* **2016**, *6*, 1339. [CrossRef]
4. Situ, C.; Mooney, M.H.; Elliott, C.T.; Buijs, J. Advances in Surface Plasmon Resonance Biosensor Technology towards High-Throughput, Food-Safety Analysis. *TrAC Trends Anal. Chem.* **2010**, *29*, 1305–1315. [CrossRef]
5. Daniyal, W.M.E.M.M.; Fen, Y.W.; Abdullah, J.; Omar, N.A.S.; Anas, N.A.A.; Ramdzan, M.; Syahira, N. Highly Sensitive Surface Plasmon Resonance Optical Sensor for Detection of Copper, Zinc, and Nickel Ions. *Sens. Lett.* **2019**, *17*, 497–504. [CrossRef]
6. Holzinger, M.; Goff, A.L.; Cosnier, S. Nanomaterials for Biosensing Applications: A Review. *Front. Chem.* **2014**, *2*, 1–10. [CrossRef]
7. Ishimaru, A.; Jaruwatanadilok, S.; Kuga, Y. Generalized Surface Plasmon Resonance Sensors Using Metamaterials and Negative Index Materials. *Prog. Electromagn. Res.* **2006**, *51*, 139–152. [CrossRef]
8. Pitarke, J.M.; Silkin, V.M.; Chulkov, E.V.; Echenique, P.M. Theory of Surface Plasmons and Surface-Plasmon Polaritons. *Rep. Prog. Phys.* **2007**, *70*, 1–87. [CrossRef]

9. Homola, J.; Yee, S.S.; Gauglitz, G. Surface Plasmon Resonance Sensors. *Sens. Actuators B Chem.* **1999**, *54*, 3–15. [CrossRef]
10. Homola, J. Surface Plasmon Resonance Sensors for Detection of Chemical and Biological Species. *Chem. Rev.* **2008**, *108*, 462–493. [CrossRef]
11. Wood, R.W. On a Remarkable Case of Uneven Distribution of Light in A Diffraction Grating Spectrum. *Proc. Phys. Soc.* **1902**, *18*, 269–275.
12. Fano, U. The Theory of Anomalous Diffraction Gratings and of Quasi-Stationary Waves on Metallic Surfaces (Sommerfeld's Waves). *JOSA* **1941**, *31*, 213–222. [CrossRef]
13. OTTO, A. Excitation of Nonradiative Surface Plasma Waves in Silver by the Method of Frustrated Total Reflection. *Z. Phys. A Hadron. Nucl.* **1968**, *216*, 398–410. [CrossRef]
14. Kretschmann, E.; Raether, H. Radiative Decay of Non Radiative Surface Plasmons Excited by Light. *Z. Naturforsch. A Phys. Sci.* **1968**, *23*, 2135–2136. [CrossRef]
15. Rizal, C. Bio-Magnetoplasmonics, Emerging Biomedical Technologies and Beyond. *J. Nanomed. Res.* **2016**, *3*. [CrossRef]
16. Fen, Y.W.; Yunus, W.M.M. Surface Plasmon Resonance Spectroscopy as An Alternative for Sensing Heavy Metal Ions: A Review. *Sens. Rev.* **2013**, *33*, 305–314. [CrossRef]
17. Homola, J.; Koudela, I.; Yee, S.S. Surface Plasmon Resonance Sensors Based on Diffraction Gratings and Prism Couplers: Sensitivity Comparison. *Sens. Actuators B Chem.* **1999**, *54*, 16–24. [CrossRef]
18. Karlssonz, R.; Fält, A. Experimental Design for Kinetic Analysis of Protein-Protein Interactions with Surface Plasmon Resonance Biosensors. *J. Immunol. Methods* **1997**, *200*, 121–133. [CrossRef]
19. Homola, J. *Electromagnetic Theory of Surface Plasmons*; Springer: Berlin/Heidelberg, Germany, 2006; pp. 3–44. [CrossRef]
20. Mukhtar, W.M.; Halim, R.M.; Hassan, H. Optimization of SPR Signals: Monitoring the Physical Structures and Refractive Indices of Prisms. *EPJ Web Conf.* **2017**, *162*. [CrossRef]
21. Al-qazwini, Y.; Noor, A.S.M.; Arasu, P.T.; Sadrolhosseini, A.R. Investigation of the Performance of an SPR-Based Optical Fiber Sensor Using Finite-Difference Time Domain. *Curr. Appl. Phys.* **2013**, *13*, 1354–1358. [CrossRef]
22. Murat, N.F.; Mukhtar, W.M.; Rashid, A.R.A.; Dasuki, K.A.; Yussuf, A.A.R.A. Optimization of Gold Thin Films Thicknesses in Enhancing SPR Response. In Proceedings of the 2016 IEEE International Conference on Semiconductor Electronics (ICSE), Kuala Lumpur, Malaysia, 17–19 August 2016; pp. 244–247. [CrossRef]
23. Michel, D.; Xiao, F.; Alameh, K. A Compact, Flexible Fiber-Optic Surface Plasmon Resonance Sensor with Changeable Sensor Chips. *Sens. Actuators B Chem.* **2017**, *246*, 258–261. [CrossRef]
24. Nelson, B.P.; Frutos, A.G.; Brockman, J.M.; Corn, R.M. Near-Infrared Surface Plasmon Resonance Measurements of Ultrathin Films. 1. Angle Shift and SPR Imaging Experiments. *Anal. Chem.* **1999**, *71*, 3928–3934. [CrossRef]
25. Patskovsky, S.; Kabashin, A.V.; Meunier, M.; Luong, J.H.T. Properties and Sensing Characteristics of Surface-Plasmon Resonance in Infrared Light. *J. Opt. Soc. Am. A* **2003**, *20*, 1644. [CrossRef] [PubMed]
26. Mukhtar, W.M.; Murat, N.F.; Samsuri, N.D.; Dasuki, K.A. Maximizing the Response of SPR Signal: A Vital Role of Light Excitation Wavelength. *AIP Conf. Proc.* **2018**, *2016*. [CrossRef]
27. Chung, J.W.; Kim, S.D.; Bernhardt, R.; Pyun, J.C. Application of SPR Biosensor for Medical Diagnostics of Human Hepatitis B Virus (HHBV). *Sens. Actuators B Chem.* **2005**, 416–422. [CrossRef]
28. Uzun, L.; Say, R.; Ünal, S.; Denizli, A. Production of Surface Plasmon Resonance Based Assay Kit for Hepatitis Diagnosis. *Biosens. Bioelectron.* **2009**, *24*, 2878–2884. [CrossRef]
29. Uludag, Y.; Tothill, I.E. Cancer Biomarker Detection in Serum Samples Using Surface Plasmon Resonance and Quartz Crystal Microbalance Sensors with Nanoparticle Signal Amplification. *Anal. Chem.* **2012**, *84*, 5898–5904. [CrossRef]
30. Ertürk, G.; Özen, H.; Tümer, M.A.; Mattiasson, B.; Denizli, A. Microcontact Imprinting Based Surface Plasmon Resonance (SPR) Biosensor for Real-Time and Ultrasensitive Detection of Prostate Specific Antigen (PSA) from Clinical Samples. *Sens. Actuators B Chem.* **2016**, *224*, 823–832. [CrossRef]
31. He, L.; Pagneux, Q.; Larroulet, I.; Serrano, A.Y.; Pesquera, A.; Zurutuza, A.; Mandler, D.; Boukherroub, R.; Szunerits, S. Label-Free Femtomolar Cancer Biomarker Detection in Human Serum Using Graphene-Coated Surface Plasmon Resonance Chips. *Biosens. Bioelectron.* **2017**, *89*, 606–611. [CrossRef]

32. Liang, R.P.; Yao, G.H.; Fan, L.X.; Qiu, J.D. Magnetic Fe 3O 4@Au Composite-Enhanced Surface Plasmon Resonance for Ultrasensitive Detection of Magnetic Nanoparticle-Enriched α-Fetoprotein. *Anal. Chim. Acta* **2012**, *737*, 22–28. [CrossRef]
33. Osman, B.; Uzun, L.; Beşirli, N.; Denizli, A. Microcontact Imprinted Surface Plasmon Resonance Sensor for Myoglobin Detection. *Mater. Sci. Eng. C* **2013**, *33*, 3609–3614. [CrossRef] [PubMed]
34. Sener, G.; Uzun, L.; Say, R.; Denizli, A. Use of Molecular Imprinted Nanoparticles as Biorecognition Element on Surface Plasmon Resonance Sensor. *Sens. Actuators B Chem.* **2011**, *160*, 791–799. [CrossRef]
35. Bocková, M.; Chadtová Song, X.; Gedeonová, E.; Levová, K.; Kalousová, M.; Zima, T.; Homola, J. Surface Plasmon Resonance Biosensor for Detection of Pregnancy Associated Plasma Protein A2 in Clinical Samples. *Anal. Bioanal. Chem.* **2016**, *408*, 7265–7269. [CrossRef] [PubMed]
36. Brun, A.P.L.; Soliakov, A.; Shah, D.S.H.; Holt, S.A.; Mcgill, A.; Lakey, J.H. Engineered Self-Assembling Monolayers for Label Free Detection of Influenza Nucleoprotein. *Biomed. Microdevices* **2015**, 49. [CrossRef]
37. Chang, Y.F.; Wang, W.H.; Hong, Y.W.; Yuan, R.Y.; Chen, K.H.; Huang, Y.W.; Lu, P.L.; Chen, Y.H.; Chen, Y.M.A.; Su, L.C.; et al. Simple Strategy for Rapid and Sensitive Detection of Avian Influenza A H7N9 Virus Based on Intensity-Modulated SPR Biosensor and New Generated Antibody. *Anal. Chem.* **2018**, *90*, 1861–1869. [CrossRef] [PubMed]
38. Zeng, C.; Huang, X.; Xu, J.; Li, G.; Ma, J.; Ji, H.F.; Zhu, S.; Chen, H. Rapid and Sensitive Detection of Maize Chlorotic Mottle Virus Using Surface Plasmon Resonance-Based Biosensor. *Anal. Biochem.* **2013**, *440*, 18–22. [CrossRef] [PubMed]
39. Cairns, T.M.; Ditto, N.T.; Atanasiu, D.; Lou, H.; Brooks, B.D.; Saw, W.T.; Eisenberg, R.J.; Cohen, G.H. Surface Plasmon Resonance Reveals Direct Binding of Herpes Simplex Virus Glycoproteins GH/GL to GD and Locates a GH/GL Binding Site on GD. *J. Virol.* **2019**, *93*, 1–21. [CrossRef] [PubMed]
40. Firdous, S.; Anwar, S.; Rafya, R. Development of Surface Plasmon Resonance (SPR) Biosensors for Use in the Diagnostics of Malignant and Infectious Diseases. *Laser Phys. Lett.* **2018**, *15*. [CrossRef]
41. Takemura, K.; Adegoke, O.; Suzuki, T.; Park, E.Y. A Localized Surface Plasmon Resonance-Amplified Immunofluorescence Biosensor for Ultrasensitive and Rapid Detection of Nonstructural Protein 1 of Zika Virus. *PLoS ONE* **2019**, *14*, 1–14. [CrossRef]
42. Yakes, B.J.; Papafragkou, E.; Conrad, S.M.; Neill, J.D.; Ridpath, J.F.; Burkhardt, W.; Kulka, M.; DeGrasse, S.L. Surface Plasmon Resonance Biosensor for Detection of Feline Calicivirus, a Surrogate for Norovirus. *Int. J. Food Microbiol.* **2013**, *162*, 152–158. [CrossRef]
43. Omar, N.A.S.; Fen, Y.W.; Abdullah, J.; Chik, C.E.N.C.E.; Mahdi, M.A. Development of an Optical Sensor Based on Surface Plasmon Resonance Phenomenon for Diagnosis of Dengue Virus E-Protein. *Sens. Bio-Sens. Res.* **2018**, *20*, 16–21. [CrossRef]
44. Omar, N.A.S.; Fen, Y.W.; Abdullah, J.; Zaid, M.H.M.; Mahdi, M.A. Structural, Optical and Sensing Properties of CdS-NH2GO Thin Film as a Dengue Virus E-Protein Sensing Material. *Optik* **2018**, *171*, 934–940. [CrossRef]
45. Omar, N.A.S.; Fen, Y.W.; Abdullah, J.; Mustapha Kamil, Y.; Daniyal, W.M.E.M.M.; Sadrolhosseini, A.R.; Mahdi, M.A. Sensitive Detection of Dengue Virus Type 2 E-Proteins Signals Using Self-Assembled Monolayers/Reduced Graphene Oxide-PAMAM Dendrimer Thin Film-SPR Optical Sensor. *Sci. Rep.* **2020**, *10*, 1–15. [CrossRef]
46. Omar, N.A.S.; Fen, Y.W.; Abdullah, J.; Sadrolhosseini, A.R.; Mustapha Kamil, Y.; Fauzi, N.I.M.; Hashim, H.S.; Mahdi, M.A. Quantitative and Selective Surface Plasmon Resonance Response Based on a Reduced Graphene Oxide–Polyamidoamine Nanocomposite for Detection of Dengue Virus E-Proteins. *Nanomaterials* **2020**, *10*, 569. [CrossRef]
47. Cenci, L.; Andreetto, E.; Vestri, A.; Bovi, M.; Barozzi, M.; Iacob, E.; Busato, M.; Castagna, A.; Girelli, D.; Bossi, A.M. Surface Plasmon Resonance Based on Molecularly Imprinted Nanoparticles for the Picomolar Detection of the Iron Regulating Hormone Hepcidin-25. *J. Nanobiotechnol.* **2015**, *13*, 1–15. [CrossRef]
48. Zhang, Q.; Jing, L.; Wang, Y.; Zhang, J.; Ren, Y.; Wang, Y.; Wei, T.; Liedberg, B. Surface Plasmon Resonance Sensor for Femtomolar Detection of Testosterone with Water-Compatible Macroporous Molecularly Imprinted Film. *Anal. Biochem.* **2014**, *463*, 7–14. [CrossRef] [PubMed]
49. Yockell-Lelièvre, H.; Bukar, N.; McKeating, K.S.; Arnaud, M.; Cosin, P.; Guo, Y.; Dupret-Carruel, J.; Mougin, B.; Masson, J.F. Plasmonic Sensors for the Competitive Detection of Testosterone. *Analyst* **2015**, *140*, 5105–5111. [CrossRef]

50. Treviño, J.; Calle, A.; Rodríguez-Frade, J.M.; Mellado, M.; Lechuga, L.M. Single- and Multi-Analyte Determination of Gonadotropic Hormones in Urine by Surface Plasmon Resonance Immunoassay. *Anal. Chim. Acta* **2009**, *647*, 202–209. [CrossRef]
51. Treviño, J.; Calle, A.; Rodríguez-Frade, J.M.; Mellado, M.; Lechuga, L.M. Surface Plasmon Resonance Immunoassay Analysis of Pituitary Hormones in Urine and Serum Samples. *Clin. Chim. Acta* **2009**, *403*, 56–62. [CrossRef]
52. Sanghera, N.; Anderson, A.; Nuar, N.; Xie, C.; Mitchell, D.; Klein-Seetharaman, J. Insulin Biosensor Development: A Case Study. *Int. J. Parallel Emerg. Distrib. Syst.* **2017**, *32*, 119–138. [CrossRef]
53. Wang, S.; Shan, X.; Patel, U.; Huang, X.; Lu, J.; Li, J.; Tao, N. Label-Free Imaging, Detection, and Mass Measurement of Single Viruses by Surface Plasmon Resonance. *Proc. Natl. Acad. Sci. USA* **2010**, *107*, 16028–16032. [CrossRef] [PubMed]
54. Masson, J.F. Surface Plasmon Resonance Clinical Biosensors for Medical Diagnostics. *ACS Sens.* **2017**, *2*, 16–30. [CrossRef] [PubMed]
55. Siedhoff, D.; Strauch, M.; Shpacovitch, V.; Merhof, D. Unsupervised Data Analysis for Virus Detection with a Surface Plasmon Resonance Sensor. In Proceedings of the 2017 Seventh International Conference on Image Processing Theory, Tools and Applications (IPTA), Montreal, QC, Canada, 28 November–1 December 2018; pp. 1–6. [CrossRef]
56. Victoria, S. Application of Surface Plasmon Resonance (SPR) for the Detection of Single Viruses and Single Biological Nano-Objects. *J. Bacteriol. Parasitol.* **2012**, *3*. [CrossRef]
57. Saylan, Y.; Yilmaz, F.; Özgür, E.; Derazshamshir, A.; Bereli, N.; Yavuz, H.; Denizli, A. Nanotechnology Characterization Tools for Biosensing and Medical Diagnosis. In *Surface Plasmon Resonance Sensors for Medical Diagnosis*; Kumar, C.S.S.R., Ed.; Springer: Berlin/Heidelberg, Germany, 2018; pp. 425–458. [CrossRef]
58. Moon, J.M.; Thapliyal, N.; Hussain, K.K.; Goyal, R.N.; Shim, Y.B. Conducting Polymer-Based Electrochemical Biosensors for Neurotransmitters: A Review. *Biosens. Bioelectron.* **2018**, *102*, 540–552. [CrossRef] [PubMed]
59. Soleymani, J. Advanced Materials for Optical Sensing and Biosensing of Neurotransmitters. *TrAC Trends Anal. Chem.* **2015**, *72*, 27–44. [CrossRef]
60. Krishna, V.M.; Somanathan, T.; Manikandan, E.; Tadi, K.K.; Uvarajan, S. Neurotransmitter Dopamine Enhanced Sensing Detection Using Fibre-Like Carbon Nanotubes by Chemical Vapor Deposition Technique. *J. Nanosci. Nanotechnol.* **2018**, *18*, 5380–5389. [CrossRef]
61. Lin, X.; Zhang, Y.; Chen, W.; Wu, P. Electrocatalytic Oxidation and Determination of Dopamine in the Presence of Ascorbic Acid and Uric Acid at a Poly (p -Nitrobenzenazo Resorcinol) Modified Glassy Carbon Electrode. *Sens. Actuators B Chem.* **2007**, *122*, 309–314. [CrossRef]
62. Liu, J.; Wang, X.; Cui, M.; Lin, L.; Jiang, S.; Jiao, L.; Zhang, L. A Promising Non-Aggregation Colorimetric Sensor of AuNRs–Ag + for Determination of Dopamine. *Sens. Actuators B Chem.* **2013**, *176*, 97–102. [CrossRef]
63. Haven, N. Dopamine Synthesis, Uptake, Metabolism, and Receptors: Relevance to Gene Therapy of Parkinson's Disease. *Exp. Neurol.* **1997**, *9*, 4–9. [CrossRef]
64. Kim, J.-H.; Auerbach, J.M.; Rodríguez-Gómez, J.A.; Velasco, I.; Gavin, D.; Lumelsky, N.; McKay, R. Dopamine neurons derived from embryonic stem cells function in an animal model of Parkinson's disease. *Nature* **2002**, *418*, 50–56. [CrossRef]
65. Pezzella, A.; Ischia, M.; Napolitano, A.; Misuraca, G.; Prota, G. Iron-Mediated Generation of the Neurotoxin 6-Hydroxydopamine Quinone by Reaction of Fatty Acid Hydroperoxides with Dopamine: A Possible Contributory Mechanism for Neuronal Degeneration in Parkinson's Disease. *J. Med. Chem.* **1997**, *40*, 2211–2216. [CrossRef] [PubMed]
66. Hyman, B.; Van Hoesen, G.; Damasio, A.; Barnes, C. Alzheimer's disease: Cell-specific pathology isolates the hippocampal formation. *Science* **1984**, *225*, 1168–1170. [CrossRef] [PubMed]
67. Wightman, M.; May, L.J.; Michael, A.C. Detection of Dopamine Dynamics in the Brain. *Anal. Chem.* **1988**, *60*, 769–779. [CrossRef]
68. Kesby, J.P. Dopamine, Psychosis and Schizophrenia: The Widening Gap between Basic and Clinical Neuroscience. *Transl. Psychiatry* **2018**, *8*, 30. [CrossRef]
69. Pandey, P.C.; Chauhan, D.S.; Singh, V. Effect of Processable Polyindole and Nanostructured Domain on the Selective Sensing of Dopamine. *Mater. Sci. Eng. C* **2012**, *32*, 1–11. [CrossRef]
70. Yu, C.; Yan, J.; Tu, Y. Electrochemiluminescent Sensing of Dopamine Using CdTe Quantum Dots Capped with Thioglycolic Acid and Supported with Carbon Nanotubes. *Microchim. Acta* **2011**, *175*, 347–354. [CrossRef]

71. Shankaran, D.R.; Iimura, K.; Kato, T. Simultaneous Determination of Ascorbic Acid and Dopamine at Sol–Gel Composite Electrode. *Sens. Actuators B Chem.* **2003**, *94*, 73–80. [CrossRef]
72. Kurzatkowska, K.; Dolusic, E.; Dehaen, W.; Sieron, K.; Radecka, H. Gold Electrode Incorporating Corrole as an Ion-Channel Mimetic Sensor for Determination of Dopamine. *Anal. Chem.* **2009**, *81*, 7397–7405. [CrossRef]
73. Lin, L.; Qiu, P.; Yang, L. Determination of Dopamine in Rat Striatum by Microdialysis and High-Performance Liquid Chromatography with Electrochemical Detection on a Functionalized Multi-Wall Carbon Nanotube Electrode. *Anal. Bioanal. Chem.* **2006**, *384*, 1308–1313. [CrossRef]
74. Zhang, L.; Lin, X. Electrochemical Behavior of a Covalently Modified Glassy Carbon Electrode with Aspartic Acid and Its Use for Voltammetric Differentiation of Dopamine and Ascorbic Acid. *Anal. Bioanal. Chem.* **2005**, *382*, 1669–1677. [CrossRef]
75. Jagadeesh, J.S.; Natarajan, S. Schizophrenia: Interaction between Dopamine, Serotonin, Glutamate, GABA. *RJPBCS* **2013**, *4*, 1267–1271.
76. Davis, K.L.; Kahn, R.S.; Ko, G.; Davidson, M. Dopamine in schizophrenia: A review and reconceptualization. *Am. J. Psychiatry* **1991**, *148*, 1474–1486. [CrossRef] [PubMed]
77. Rui, Z.; Huang, W.; Chen, Y.; Zhang, K.; Cao, Y.; Tu, J. Facile Synthesis of Graphene / Polypyrrole 3D Composite for a High-Sensitivity Non-Enzymatic Dopamine Detection. *J. Appl. Polym. Sci.* **2017**, *134*, 44840. [CrossRef]
78. Roy, A.; Pickar, D.; De Jong, J.; Karoum, F.; Linnoila, M. Norepinephrine and its metabolites in cerebrospinal fluid, plasma, and urine: Relationship to hypothalamic-pituitary-adrenal axis function in depression. *Arch. Gen. Psychiatry* **1988**, *45*, 849–857. [CrossRef] [PubMed]
79. Okumura, T.; Nakajima, Y.; Matsuoka, M.; Takamatsu, T. Study of Salivary Catecholamines Using Fully Automated Column-Switching High-Performance Liquid Chromatography. *J. Chromatogr. B Biomed. Appl.* **1997**, *694*, 305–316. [CrossRef]
80. Cao, Y.; Mcdermott, M.T. Femtomolar and Selective Dopamine Detection by a Gold Nanoparticle Enhanced Surface Plasmon Resonance Aptasensor. *bioRxiv* **2018**, 1–24. [CrossRef]
81. Yoshitake, T.; Yoshitake, S.; Fujino, K.; Nohta, H.; Yamaguchi, M.; Kehr, J. High-Sensitive Liquid Chromatographic Method for Determination of Neuronal Release of Serotonin, Noradrenaline and Dopamine Monitored by Microdialysis in the Rat Prefrontal Cortex. *J. Neurosci. Methods* **2004**, *140*, 163–168. [CrossRef]
82. Carrera, V.; Sabater, E.; Vilanova, E.; Sogorb, M.A. A Simple and Rapid HPLC-MS Method for the Simultaneous Determination of Epinephrine, Norepinephrine, Dopamine and 5-Hydroxytryptamine: Application to the Secretion of Bovine Chromaffin Cell Cultures. *J. Chromatogr. B Anal. Technol. Biomed. Life Sci.* **2007**, *847*, 88–94. [CrossRef]
83. Muzzi, C.; Bertocci, E.; Terzuoli, L.; Porcelli, B.; Ciari, I.; Pagani, R.; Guerranti, R. Simultaneous Determination of Serum Concentrations of Levodopa, Dopamine, 3-O-Methyldopa and α-Methyldopa by HPLC. *Biomed. Pharmacother.* **2008**, *62*, 253–258. [CrossRef]
84. Woolley, A.T.; Lao, K.; Glazer, A.N.; Mathies, R.A. Capillary Electrophoresis Chips with Integrated Electrochemical Detection. *Anal. Chem.* **1998**, *70*, 684–688. [CrossRef]
85. Wang, L.; Liu, Y.; Xie, H.; Fu, Z. Trivalent Copper Chelate-Luminol Chemiluminescence System for Highly Sensitive CE Detection of Dopamine in Biological Sample after Clean up Using SPE. *Electrophoresis* **2012**, *33*, 1589–1594. [CrossRef] [PubMed]
86. Zhao, Y.; Zhao, S.; Huang, J.; Ye, F. Quantum Dot-Enhanced Chemiluminescence Detection for Simultaneous Determination of Dopamine and Epinephrine by Capillary Electrophoresis. *Talanta* **2011**, *85*, 2650–2654. [CrossRef] [PubMed]
87. Thabano, J.R.E.; Breadmore, M.C.; Hutchinson, J.P.; Johns, C.; Haddad, P.R. Silica Nanoparticle-Templated Methacrylic Acid Monoliths for in-Line Solid-Phase Extraction-Capillary Electrophoresis of Basic Analytes. *J. Chromatogr. A* **2009**, *1216*, 4933–4940. [CrossRef] [PubMed]
88. Wang, X.; Jin, B.; Lin, X. In-Situ FTIR Spectroelectrochemical Study of Dopamine at a Glassy Carbon Electrode in a Neutral Solution. *Anal. Sci.* **2002**, *18*, 931–933. [CrossRef]
89. Abaidur, S.M.; Alothman, Z.A.; Alam, S.M.; Lee, S.H. Flow Injection-Chemiluminescence Determination of Dopamine Using Potassium Permanganate and Formaldehyde System. *Spectrochim. Acta Part A Mol. Biomol. Spectrosc.* **2012**, *96*, 221–225. [CrossRef]

90. Fritzen-Garcia, M.B.; Monteiro, F.F.; Cristofolini, T.; Acuña, J.J.S.; Zanetti-Ramos, B.G.; Oliveira, I.R.W.Z.; Soldi, V.; Pasa, A.A.; Creczynski-Pasa, T.B. Characterization of Horseradish Peroxidase Immobilized on PEGylated Polyurethane Nanoparticles and Its Application for Dopamine Detection. *Sens. Actuators B Chem.* **2013**, *182*, 264–272. [CrossRef]
91. Liu, S.; Sun, W.; Hu, F. Graphene Nano Sheet-Fabricated Electrochemical Sensor for the Determination of Dopamine in the Presence of Ascorbic Acid Using Cetyltrimethylammonium Bromide as the Discriminating Agent. *Sens. Actuators B Chem.* **2012**, *173*, 497–504. [CrossRef]
92. Sajid, M.; Nazal, M.K.; Mansha, M.; Alsharaa, A.; Jillani, S.M.S.; Basheer, C. Chemically Modified Electrodes for Electrochemical Detection of Dopamine in the Presence of Uric Acid and Ascorbic Acid: A Review. *TrAC Trends Anal. Chem.* **2016**, *76*, 15–29. [CrossRef]
93. Shin, J.-W.; Kim, K.-J.; Yoon, J.; Jo, J.; El-Said, W.A.; Choi, J.-W. Silver Nanoparticle Modified Electrode Covered by Graphene Oxide for the Enhanced Electrochemical Detection of Dopamine. *Sensors* **2017**, *17*, 2771. [CrossRef]
94. Hows, M.E.P.; Lacroix, L.; Heidbreder, C.; Organ, A.J.; Shah, A.J. High-Performance Liquid Chromatography/Tandem Mass Spectrometric Assay for the Simultaneous Measurement of Dopamine, Norepinephrine, 5-Hydroxytryptamine and Cocaine in Biological Samples. *J. Neurosci. Methods* **2004**, *138*, 123–132. [CrossRef]
95. Moini, M.; Schultz, C.L.; Mahmood, H. CE/Electrospray Ionization-MS Analysis of Underivatized D/L-Amino Acids and Several Small Neurotransmitters at Attomole Levels through the Use of 18-Crown-6-Tetracarboxylic Acid as a Complexation Reagent/Background Electrolyte. *Anal. Chem.* **2003**, *75*, 6282–6287. [CrossRef] [PubMed]
96. Syslová, K.; Rambousek, L.; Kuzma, M.; Najmanová, V.; Bubeníková-Valešová, V.; Šlamberová, R.; Kačer, P. Monitoring of Dopamine and Its Metabolites in Brain Microdialysates: Method Combining Freeze-Drying with Liquid Chromatography-Tandem Mass Spectrometry. *J. Chromatogr. A* **2011**, *1218*, 3382–3391. [CrossRef] [PubMed]
97. Reza Hormozi Nezhad, M.; Tashkhourian, J.; Khodaveisi, J.; Reza Khoshi, M. Simultaneous Colorimetric Determination of Dopamine and Ascorbic Acid Based on the Surface Plasmon Resonance Band of Colloidal Silver Nanoparticles Using Artificial Neural Networks. *Anal. Methods* **2010**, *2*, 1263–1269. [CrossRef]
98. Wang, H.Y.; Hui, Q.S.; Xu, L.X.; Jiang, J.G.; Sun, Y. Fluorimetric Determination of Dopamine in Pharmaceutical Products and Urine Using Ethylene Diamine as the Fluorigenic Reagent. *Anal. Chim. Acta* **2003**, *497*, 93–99. [CrossRef]
99. Kruss, S.; Landry, M.P.; Vander Ende, E.; Lima, B.M.A.; Reuel, N.F.; Zhang, J.; Nelson, J.; Mu, B.; Hilmer, A.; Strano, M. Neurotransmitter Detection Using Corona Phase Molecular Recognition on Fluorescent Single-Walled Carbon Nanotube Sensors. *J. Am. Chem. Soc.* **2014**, *136*, 713–724. [CrossRef] [PubMed]
100. Zhao, F.; Kim, J. Fabrication of a Dopamine Sensor Based on Carboxyl Quantum Dots. *J. Nanosci. Nanotechnol.* **2015**, *15*, 7871–7875. [CrossRef] [PubMed]
101. Kruss, S.; Salem, D.P.; Vuković, L.; Lima, B.; Vander Ende, E.; Boyden, E.S.; Strano, M.S. High-Resolution Imaging of Cellular Dopamine Efflux Using a Fluorescent Nanosensor Array. *Proc. Natl. Acad. Sci. USA* **2017**, *114*, 1789–1794. [CrossRef]
102. Qi, H.; Peng, Y.; Gao, Q.; Zhang, C. Applications of Nanomaterials in Electrogenerated Chemiluminescence Biosensors. *Sensors* **2009**, *9*, 674–695. [CrossRef]
103. Bu, Y.; Lee, S. Influence of Dopamine Concentration and Surface Coverage of Au Shell on the Optical Properties of Au, Ag, and Ag CoreAu Shell Nanoparticles. *ACS Appl. Mater. Interfaces* **2012**, *4*, 3923–3931. [CrossRef]
104. Bu, Y.; Lee, S.-W. Optical Properties of Dopamine Molecules with Silver Nanoparticles as Surface-Enhanced Raman Scattering (SERS) Substrates at Different PH Conditions. *J. Nanosci. Nanotechnol.* **2013**, *13*, 5992–5996. [CrossRef]
105. Ranc, V.; Markova, Z.; Hajduch, M.; Prucek, R.; Kvitek, L.; Kaslik, J.; Safarova, K.; Zboril, R. Magnetically Assisted Surface-Enhanced Raman Scattering Selective Determination of Dopamine in an Artificial Cerebrospinal Fluid and a Mouse Striatum Using Fe_3O_4/Ag Nanocomposite. *Anal. Chem.* **2014**, *86*, 2939–2946. [CrossRef] [PubMed]

106. An, J.H.; Choi, D.K.; Lee, K.J.; Choi, J.W. Surface-Enhanced Raman Spectroscopy Detection of Dopamine by DNA Targeting Amplification Assay in Parkisons's Model. *Biosens. Bioelectron.* **2015**, *67*, 739–746. [CrossRef] [PubMed]
107. Wang, P.; Xia, M.; Liang, O.; Sun, K.; Cipriano, A.F.; Schroeder, T.; Liu, H.; Xie, Y.H. Label-Free SERS Selective Detection of Dopamine and Serotonin Using Graphene-Au Nanopyramid Heterostructure. *Anal. Chem.* **2015**, *87*, 10255–10261. [CrossRef] [PubMed]
108. Lu, J.; Xu, C.; Nan, H.; Zhu, Q.; Qin, F.; Manohari, A.G.; Wei, M.; Zhu, Z.; Shi, Z.; Ni, Z. SERS-Active ZnO/Ag Hybrid WGM Microcavity for Ultrasensitive Dopamine Detection. *Appl. Phys. Lett.* **2016**, *109*. [CrossRef]
109. Deftereos, N.T.; Calokerinos, A.C.; Efstathiou, C.E. Flow Injection Chemiluminometric Determination of Epinephrine, Norepinephrine, Dopamine and L-DOPA. *Analyst* **1993**, *118*, 627–632. [CrossRef]
110. Dutta, P.; Pernites, R.B.; Danda, C.; Advincula, R.C. SPR Detection of Dopamine Using Cathodically Electropolymerized, Molecularly Imprinted Poly-p-Aminostyrene Thin Films. *Macromol. Chem. Phys.* **2011**, *212*, 2439–2451. [CrossRef]
111. Jia, K.; Khaywah, M.Y.; Li, Y.; Bijeon, J.L.; Adam, P.M.; Déturche, R.; Guelorget, B.; François, M.; Louarn, G.; Ionescu, R.E. Strong Improvements of Localized Surface Plasmon Resonance Sensitivity by Using Au/Ag Bimetallic Nanostructures Modified with Polydopamine Films. *ACS Appl. Mater. Interfaces* **2014**, *6*, 219–227. [CrossRef]
112. Sebők, D.; Csapó, E.; Preočanin, T.; Bohus, G.; Kallay, N.; Dékány, I. Adsorption of Ibuprofen and Dopamine on Functionalized Gold Using Surface Plasmon Resonance Spectroscopy at Solid-Liquid Interface. *Croat. Chem. Acta* **2013**, *86*, 287–295. [CrossRef]
113. Choi, J.-H.; Lee, J.-H.; Oh, B.-K.; Choi, J.-W. Localized Surface Plasmon Resonance-Based Label-Free Biosensor for Highly Sensitive Detection of Dopamine. *J. Nanosci. Nanotechnol.* **2014**, *14*, 5658–5661. [CrossRef]
114. Su, R.; Pei, Z.; Huang, R.; Qi, W.; Wang, M.; Wang, L.; He, Z. Polydopamine-Assisted Fabrication of FiberOptic Localized Surface Plasmon Resonance Sensor Based on Gold Nanoparticles. *Trans. Tianjin Univ.* **2015**, *21*, 412–419. [CrossRef]
115. Kamal Eddin, F.B.; Fen, Y.W. Recent Advances in Electrochemical and Optical Sensing of Dopamine. *Sensors* **2020**, *20*, 1039. [CrossRef] [PubMed]
116. Omar, N.A.S.; Fen, Y.W. Recent Development of SPR Spectroscopy as Potential Method for Diagnosis of Dengue Virus E-Protein. *Sens. Rev.* **2018**, *38*, 106–116. [CrossRef]
117. Omar, N.A.S.; Fen, Y.W.; Saleviter, S.; Daniyal, W.M.E.M.M.; Anas, N.A.A.; Ramdzan, N.S.M.; Roshidi, M.D.A. Development of a Graphene-Based Surface Plasmon Resonance Optical Sensor Chip for Potential Biomedical Application. *Materials* **2019**, *12*, 1928. [CrossRef] [PubMed]
118. Zainuddin, N.H.; Fen, Y.W.; Alwahib, A.A.; Yaacob, M.H.; Bidin, N.; Omar, N.A.S.; Mahdi, M.A. Detection of Adulterated Honey by Surface Plasmon Resonance Optical Sensor. *Optik* **2018**, *168*, 134–139. [CrossRef]
119. Sadrolhosseini, A.R.; Rashid, S.A.; Jamaludin, N.; Noor, A.S.M. Surface Plasmon Resonance Sensor Using Polypyrrole-Chitosan/Graphene Quantum Dots Layer for Detection of Sugar. *Mater. Res. Express* **2019**, *6*, 075028. [CrossRef]
120. Roshidi, M.D.A.; Fen, Y.W.; Daniyal, W.M.E.M.M.; Omar, N.A.S.; Zulholinda, M. Structural and Optical Properties of Chitosan–Poly(Amidoamine) Dendrimer Composite Thin Film for Potential Sensing Pb^{2+} Using an Optical Spectroscopy. *Optik* **2019**, *185*, 351–358. [CrossRef]
121. Daniyal, W.M.E.M.M.; Fen, Y.W.; Abdullah, J.; Sadrolhosseini, A.R.; Saleviter, S.; Omar, N.A.S. Label-Free Optical Spectroscopy for Characterizing Binding Properties of Highly Sensitive Nanocrystalline Cellulose-Graphene Oxide Based Nanocomposite towards Nickel Ion. *Spectrochim. Acta A* **2019**, *212*, 25–31. [CrossRef]
122. Roshidi, M.D.A.; Fen, Y.W.; Omar, N.A.S.; Saleviter, S.; Daniyal, W.M.E.M.M. Optical Studies of Graphene Oxide/Poly(Amidoamine) Dendrimer Composite Thin Film and Its Potential for Sensing Hg^{2+} Using Surface Plasmon Resonance Spectroscopy. *Sens. Mater.* **2019**, *31*, 1147. [CrossRef]
123. Zainudin, A.A.; Fen, Y.W.; Yusof, N.A.; Al-Rekabi, S.H.; Mahdi, M.A.; Omar, N.A.S. Incorporation of Surface Plasmon Resonance with Novel Valinomycin Doped Chitosan-Graphene Oxide Thin Film for Sensing Potassium Ion. *Spectrochim. Acta A* **2018**, *191*, 111–115. [CrossRef]
124. Sadrolhosseini, A.R.; Naseri, M.; Rashid, S.A. Polypyrrole-Chitosan/Nickel-Ferrite Nanoparticle Composite Layer for Detecting Heavy Metal Ions Using Surface Plasmon Resonance Technique. *Opt. Laser Technol.* **2017**, *93*, 216–223. [CrossRef]

125. Alwahib, A.A.; Sadrolhosseini, A.R.; An'Amt, M.N.; Lim, H.N.; Yaacob, M.H.; Abu Bakar, M.H.; Ming, H.N.; Mahdi, M.A. Reduced Graphene Oxide/Maghemite Nanocomposite for Detection of Hydrocarbon Vapor Using Surface Plasmon Resonance. *IEEE Photonics J.* **2016**, *8*, 1–9. [CrossRef]
126. Fen, Y.W.; Yunus, W.M.M.; Talib, Z.A.; Yusof, N.A. Development of Surface Plasmon Resonance Sensor for Determining Zinc Ion Using Novel Active Nanolayers as Probe. *Spectrochim. Acta A* **2015**, *134*, 48–52. [CrossRef] [PubMed]
127. Matsui, J.; Akamatsu, K.; Hara, N.; Miyoshi, D.; Nawafune, H.; Tamaki, K.; Sugimoto, N. SPR Sensor Chip for Detection of Small Molecules Using Molecularly Imprinted Polymer with Embedded Gold Nanoparticles. *Anal. Chem.* **2005**, *77*, 4282–4285. [CrossRef] [PubMed]
128. Matsui, J.; Akamatsu, K.; Nishiguchi, S.; Miyoshi, D.; Nawafune, H.; Tamaki, K.; Sugimoto, N. Composite of Au Nanoparticles and Molecularly Imprinted Polymer as a Sensing Material. *Anal. Chem.* **2004**, *76*, 1310–1315. [CrossRef] [PubMed]
129. Kumbhat, S.; Shankaran, D.R.; Kim, S.J.; Gobi, K.V.; Joshi, V.; Miura, N. Surface Plasmon Resonance Biosensor for Dopamine Using D3 Dopamine Receptor as a Biorecognition Molecule. *Biosens. Bioelectron.* **2007**, *23*, 421–427. [CrossRef] [PubMed]
130. Zangeneh Kamali, K.; Pandikumar, A.; Sivaraman, G.; Lim, H.N.; Wren, S.P.; Sun, T.; Huang, N.M. Silver@graphene Oxide Nanocomposite-Based Optical Sensor Platform for Biomolecules. *RSC Adv.* **2015**, *5*, 17809–17816. [CrossRef]
131. Rithesh Raj, D.; Prasanth, S.; Vineeshkumar, T.V.; Sudarsanakumar, C. Surface Plasmon Resonance Based Fiber Optic Dopamine Sensor Using Green Synthesized Silver Nanoparticles. *Sens. Actuators B Chem.* **2016**, *224*, 600–606. [CrossRef]
132. Jiang, K.; Wang, Y.; Thakur, G.; Kotsuchibashi, Y.; Naicker, S.; Narain, R.; Thundat, T. Rapid and Highly Sensitive Detection of Dopamine Using Conjugated Oxaborole-Based Polymer and Glycopolymer Systems. *ACS Appl. Mater. Interfaces* **2017**, *9*, 15225–15231. [CrossRef]
133. Park, S.J.; Lee, S.H.; Yang, H.; Park, C.S.; Lee, C.S.; Kwon, O.S.; Park, T.H.; Jang, J. Human Dopamine Receptor-Conjugated Multidimensional Conducting Polymer Nanofiber Membrane for Dopamine Detection. *ACS Appl. Mater. Interfaces* **2016**, *8*, 28897–28903. [CrossRef]
134. Manaf, A.A.; Ghadiry, M.; Soltanian, R.; Ahmad, H.; Lai, C.K. Picomole Dopamine Detection Using Optical Chips. *Plasmonics* **2017**, *12*, 1505–1510. [CrossRef]
135. Sharma, S.; Gupta, B.D. Surface Plasmon Resonance Based Highly Selective Fiber Optic Dopamine Sensor Fabricated Using Molecular Imprinted GNP/SnO$_2$ Nanocomposite. *J. Light. Technol.* **2018**, *36*, 5956–5962. [CrossRef]
136. Sun, J.; Jiang, S.; Xu, J.; Li, Z.; Li, C.; Jing, Y.; Zhao, X.; Pan, J.; Zhang, C. and Man, B. Sensitive and Selective SPR Sensor Employing Gold-Supported Graphene Composite Film/D-Shaped Fiber for Dopamine Detection. *J. Phys. D Appl. Phys.* **2019**, *29*, 465705. [CrossRef]
137. Yuan, Y.J.; Xu, Z.; Chen, Y. Investigation of Dopamine Immobilized on Gold by Surface Plasmon Resonance. *AIP Adv.* **2019**, *9*, 035028. [CrossRef]
138. Wang, W.; Wang, W.; Davis, J.J.; Luo, X. Ultrasensitive and Selective Voltammetric Aptasensor for Dopamine Based on a Conducting Polymer Nanocomposite Doped with Graphene Oxide. *Microchim. Acta* **2015**, *182*, 1123–1129. [CrossRef]
139. Zhang, Z.F.; Cui, H.; Lai, C.Z.; Liu, L.J. Gold Nanoparticle-Catalyzed Luminol Chemiluminescence and Its Analytical Applications. *Anal. Chem.* **2005**, *77*, 3324–3329. [CrossRef]
140. Li, Q.; Zheng, J.Y.; Yan, Y.; Zhao, Y.S.; Yao, J. Electrogenerated Chemiluminescence of Metal-Organic Complex Nanowires: Reduced Graphene Oxide Enhancement and Biosensing Application. *Adv. Mater.* **2012**, *24*, 4745–4749. [CrossRef]
141. Liu, X.; Hu, X.; Xie, Z.; Chen, P.; Sun, X.; Yan, J.; Zhou, S. In Situ Bifunctionalized Carbon Dots with Boronic Acid and Amino Groups for Ultrasensitive Dopamine Detection. *Anal. Methods* **2016**, *8*, 3236–3241. [CrossRef]
142. Liang, W.; He, S.; Fang, J. Self-Assembly of J-Aggregate Nanotubes and Their Applications for Sensing Dopamine. *Langmuir* **2014**, *30*, 805–811. [CrossRef]
143. Fang, X.; Ren, H.; Zhao, H.; Li, Z. Ultrasensitive Visual and Colorimetric Determination of Dopamine Based on the Prevention of Etching of Silver Nanoprisms by Chloride. *Microchim. Acta* **2017**, *184*, 415–421. [CrossRef]

144. Tang, L.; Li, S.; Han, F.; Liu, L.; Xu, L.; Ma, W.; Kuang, H.; Li, A.; Wang, L.; Xu, C. SERS-Active Au@Ag Nanorod Dimers for Ultrasensitive Dopamine Detection. *Biosens. Bioelectron.* **2015**, *71*, 7–12. [CrossRef]
145. Dong, J.X.; Li, N.B.; Luo, H.Q. The Formation of Zirconium Hexacyanoferrate(II) Nanoparticles and Their Application in the Highly Sensitive Determination of Dopamine Based on Enhanced Resonance Rayleigh Scattering. *Anal. Methods* **2013**, *5*, 5541–5548. [CrossRef]
146. Qin, W.W.; Wang, S.P.; Li, J.; Peng, T.H.; Xu, Y.; Wang, K.; Shi, J.Y.; Fan, C.H.; Li, D. Visualizing Dopamine Released from Living Cells Using a Nanoplasmonic Probe. *Nanoscale* **2015**, *7*, 15070–15074. [CrossRef] [PubMed]
147. Taghdiri, M.; Mohamadipour-taziyan, A. Application of Sephadex LH-20 for Microdetermination of Dopamine by Solid Phase Spectrophotometry. *ISRN Pharm.* **2012**, *2012*, 1–5. [CrossRef] [PubMed]
148. Sun, B.; Wang, C. High-Sensitive Sensor of Dopamine Based on Photoluminescence Quenching of Hierarchical CdS Spherical Aggregates. *J. Nanomater.* **2012**, *2012*. [CrossRef]
149. Zeng, Z.; Cui, B.; Wang, Y.; Sun, C.; Zhao, X.; Cui, H. Dual Reaction-Based Multimodal Assay for Dopamine with High Sensitivity and Selectivity Using Functionalized Gold Nanoparticles. *ACS Appl. Mater. Interfaces* **2015**, *7*, 16518–16524. [CrossRef]
150. Hun, X.; Wang, S.; Wang, S.; Zhao, J.; Luo, X. A Photoelectrochemical Sensor for Ultrasensitive Dopamine Detection Based on Single-Layer NanoMoS$_2$ Modified Gold Electrode. *Sens. Actuators B Chem.* **2017**, *249*, 83–89. [CrossRef]
151. Helmerhorst, E.; Chandler, D.J.; Nussio, M.; Mamotte, C.D. Real-Time and Label-Free Bio-Sensing of Molecular Interactions by Surface Plasmon Resonance: A Laboratory Medicine Perspective. *Clin. Biochem. Rev.* **2012**, *33*, 161–173.
152. Li, Z.Y.; Xia, Y. Metal Nanoparticles with Gain toward Single-Molecule Detection by Surface-Enhanced Raman Scattering. *Nano Lett.* **2010**, *10*, 243–249. [CrossRef]

© 2020 by the authors. Licensee MDPI, Basel, Switzerland. This article is an open access article distributed under the terms and conditions of the Creative Commons Attribution (CC BY) license (http://creativecommons.org/licenses/by/4.0/).

MDPI
St. Alban-Anlage 66
4052 Basel
Switzerland
www.mdpi.com

Molecules Editorial Office
E-mail: molecules@mdpi.com
www.mdpi.com/journal/molecules

Disclaimer/Publisher's Note: The statements, opinions and data contained in all publications are solely those of the individual author(s) and contributor(s) and not of MDPI and/or the editor(s). MDPI and/or the editor(s) disclaim responsibility for any injury to people or property resulting from any ideas, methods, instructions or products referred to in the content.

www.ingramcontent.com/pod-product-compliance
Lightning Source LLC
LaVergne TN
LVHW070613100526
838202LV00012B/641